Benjamin Osgood Peirce

Elements of the theory of the Newtonian potential function

Benjamin Osgood Peirce

Elements of the theory of the Newtonian potential function

ISBN/EAN: 9783337277987

Printed in Europe, USA, Canada, Australia, Japan

Cover: Foto ©berggeist007 / pixelio.de

More available books at **www.hansebooks.com**

ELEMENTS

OF THE

THEORY OF THE NEWTONIAN POTENTIAL FUNCTION.

BY

B. O. PEIRCE, Ph.D.,

ASSISTANT PROFESSOR OF MATHEMATICS AND PHYSICS
IN HARVARD UNIVERSITY.

BOSTON:
PUBLISHED BY GINN & COMPANY.
1886.

J. S. CUSHING & Co., PRINTERS, BOSTON.

PREFACE.

———❖———

THIS book is almost entirely made up of lecture-notes which from time to time during the last four years I have written out for the use of students who have begun with me the study of what I have ventured to call, after Neumann, the Newtonian Potential Function.

The notes were intended for readers somewhat familiar with the principles of the Differential and Integral Calculus, but unacquainted with many of the methods commonly used in applying Mathematics to the study of physical problems. These students, I learned, found it difficult to get from any single book in English a treatment of the subject at once elementary enough to be within their easy comprehension, and at the same time suited to the purposes of such of them as intended eventually to pursue the subject farther, or wished, without necessarily making a specialty of Mathematical Physics, to prepare themselves to study Experimental Physics thoroughly and understandingly. What is here printed seems to have been of use to some of those who have read it in manuscript, and it is hoped that it may now be helpful to a larger number of students.

Since these notes are professedly elementary in character, I feel that no apology is needed for what may seem to be the rather prolix way in which some of the subjects are treated, or for an arrangement of matter which would be

unsuitable in a book intended for a different class of readers.
I have not hesitated to use a long proof whenever this has
seemed to me more easily comprehensible than a short and
mathematically neater one, and I have often given more than
one demonstration of a single theorem in order to illustrate
different methods of work.   Although I have used freely
the notation* of the Calculus, I have assumed on the part
of the reader only an elementary knowledge of its principles.

The short treatment of Electrostatics in Chapter v. is introduced to show how the theorems of the preceding chapters
may be used in solving physical problems; but it is hoped that
a person who has mastered even the little here given will be
able to understand, with the aid of some good treatise on
Experimental Physics, most of the phenomena of Electrostatics.   It is also hoped that those readers who mean to study
the subject of Electricity from the mathematical point of

---

* In this book the change made in a function u by giving to the
independent variable x the arbitrary, finite increment Δx, and keeping the
other independent variables, if there are any, unchanged, is denoted by
$\Delta_x u$. Similarly, $\Delta_y u$ and $\Delta_z u$ express the increments of u due to changes
respectively in y alone and in z alone. The total change in u due to
simultaneous changes in all the independent variables is sometimes
denoted by Δu; so that if $u = f(x, y, z)$,

$$\Delta u = \frac{\Delta_x u}{\Delta x} \cdot \Delta x + \frac{\Delta_y u}{\Delta y} \cdot \Delta y + \frac{\Delta_z u}{\Delta z} \cdot \Delta z + \epsilon,$$

where ϵ is an infinitesimal of an order higher than the first.

The partial derivatives of u with respect to x, y, and z are denoted by
$D_x u$, $D_y u$, and $D_z u$, and the sign $\doteq$ placed between a variable and a constant is used to show that the former is to be made to approach the
latter as its limit. In those cases where it is desirable to draw attention
to the fact that a certain derivative is total, the differential notation
$\frac{du}{dx}$ is used.

view will find what they have learned here useful when they take up standard works on the subject.

My sincere thanks are due to H. N. Wheeler, A.M., who has read much of the manuscript of the following pages and all of the proof, and to Dr. E. H. Hall, who has examined parts of Chapters iv. and v. and helped me with various suggestions. I am indebted to other friends also, and among them to Mr. W. A. Stone for the use of some of his problems.

The reader who wishes to get a thorough knowledge of the properties of the Newtonian Potential Function and of its applications to problems in Electricity is referred to the following works, which, with others, I have consulted and used in writing these notes.

Betti: Teorica delle Forze Newtoniane e sue Applicazioni all' Elettrostatica e al Magnetismo.

Clausius: Die Potentialfunction und das Potential.

Cumming: An Introduction to the Theory of Electricity.

Chrystal: The article "Electricity" in the Ninth Edition of the Encyclopædia Britannica.

Dirichlet: Vorlesungen über die im umgekehrten Verhältniss des Quadrats der Entfernung wirkenden Kräfte.

Gauss: Allgemeine Lehrsätze in Beziehung auf die im verkehrten Verhältnisse des Quadrates der Entfernung wirkenden Anziehungs- und Abstossungskräfte. Also other papers to be found in Volume V. of his Gesammelte Werke.

Green: An Essay on the Application of Mathematical Analysis to the Theories of Electricity and Magnetism.*

Mascart: Traité d'Electricité Statique. Also Wallentin's translation of the same work into German, with additions.

---

* A copy of the original edition of this paper is to be found in the Library of Harvard University, Gore Hall, Cambridge. The paper has been reprinted by Ferrers in "The Mathematical Papers of George Green," and by Thomson in Crelle's Journal.

Mascart et Joubert: Leçons sur l'Electricité et le Magnetisme. Also Atkinson's translation of the same work into English, with additions.

Mathieu: Théorie du Potential et ses Applications à l'Electrostatique et au Magnétisme.

Maxwell: An Elementary Treatise on Electricity. A Treatise on Electricity and Magnetism.

Minchin: A Treatise on Statics.

C. Neumann: Untersuchungen über das Logarithmische und Newton'sche Potential.

Riemann: Schwere, Electricität und Magnetismus, edited by Hattendorff.

Schell: Theorie der Bewegung und der Kräfte.

Thomson: Reprint of Papers on Electrostatics and Magnetism.

Thomson and Tait: A Treatise on Natural Philosophy.

Todhunter: A Treatise on Analytical Statics.

Watson and Burbury: The Mathematical Theory of Electricity and Magnetism.

Wiedemann: Die Lehre von der Electricität.

TABLE OF CONTENTS.

CHAPTER I.

THE ATTRACTION OF GRAVITATION.

SECTION.		PAGE
1.	The law of gravitation	1
2.	The attraction at a point	1
3.	The unit of force	2
4.	The attraction due to discrete particles	2
5.	The attraction of a straight wire at a point in its axis	3
6.	The attraction at any point due to a straight wire	4
7.	The attraction at a point in its axis due to a cylinder of revolution	7
8.	The attraction at the vertex of a cone of revolution due to the whole cone and to different frusta	8
9.	The attraction due to a homogeneous spherical shell; to a solid sphere	11
10.	The attraction due to a homogeneous hemisphere	13
11.	Apparent anomalies in the latitudes of places near the foot of a hemispherical hill	15
12.	The attraction due to any ellipsoidal homoeoid is zero at all points within the cavity enclosed by the shell	16
13.	The attraction due to a spherical shell whose density at any point depends upon the distance of the point from the centre	18
14.	The attraction at any point due to any given mass	19
15.	The component in any direction of the attraction at a point P due to a given mass is always finite	21

SECTION   PAGE
16. The attraction between two straight wires 22

17. The attraction between two spheres 23

18. The attraction between any two rigid bodies .   .   . 24

CHAPTER II.

THE NEWTONIAN POTENTIAL FUNCTION IN THE CASE OF GRAVITATION.

19. Definition of the potential function 29

20. The derivatives of the potential function relative to the space coördinates are functions of these coördinates which represent the components parallel to the coördinate axes of the attraction at the point (x, y, z) . 30

21. Extension of the statement of the last section .   .   . 31

22. The potential function due to a given attracting mass is everywhere finite, and the statements of the two preceding sections hold good for points within the attracting mass 32

23. The potential function due to a straight wire .   .   . 34

24. The potential function due to a spherical shell .   .   . 35

25. Equipotential surfaces and their properties .   .   . 37

26. The potential function is zero at infinity .   .   . 40

27. The potential function as a measure of work .   .   . 40

28. Laplace's Equation 42

29. The second derivatives of the potential function are finite at points within the attracting mass 43

30. The first derivatives of the potential function change continuously as the point. (x, y, z) moves through the boundaries of an attracting mass 48

31. Theorem due to Gauss. The potential function can have no maxima or minima at points of empty space .   . 50

32. Tubes of force and their properties 53

33. Spherical distributions of matter and their attractions   . 54

34. Cylindrical distributions of matter and their attractions   . 58

SECTION   PAGE

35. Poisson's Equation obtained by the application of Gauss's Theorem to volume elements 59

36. Poisson's Equation in the integral form 62

37. The average value of the potential function on a spherical surface 64

38. The equilibrium of fluids at rest under the action of given forces 66

CHAPTER III.

THE NEWTONIAN POTENTIAL FUNCTION IN THE CASE OF REPULSION.

39. Repulsion according to the "Law of Nature" . . . 72

40. The force at any point due to a given distribution of repelling matter 73

41. The potential function due to repelling matter as a measure of work 75

42. Gauss's Theorem in the case of repelling matter . . . 75

43. Poisson's Equation in the case of repelling matter . . 76

44. The coexistence of two kinds of active matter . . . 77

CHAPTER IV.

THE PROPERTIES OF SURFACE DISTRIBUTIONS. GREEN'S THEOREM.

45. The force due to a closed shell of repelling matter . . 80

46. The potential function is finite at points in a surface distribution of matter 82

47. The normal force at any point of a surface distribution . 85

48. Green's Theorem 87

49. Special cases under Green's Theorem 91

50. Surface distributions which are equivalent to certain volume distributions 95

51. Those characteristics of the potential function which are sufficient to determine the function . . . 96

52. Thomson's Theorem. Dirichlet's Principle . . . 98

CHAPTER V.

ELECTROSTATICS.

SECTION   PAGE

53. Introductory 103

54. The charges on conductors are superficial 104

55. General principles which follow directly from the theory
    of the Newtonian potential function 106

56. Tubes of force and their properties 108

57. Hollow conductors 110

58. The charge induced on a conductor which is put to earth . 114

59. Coefficients of induction and capacity 115

60. The distribution of electricity on a spherical conductor . 117

61. The distribution of a given charge on an ellipsoidal conductor 118

62. Spherical condensers 119

63. Condensers made of two parallel conducting plates . . 122

64. The capacity of a long cylinder surrounded by a concentric
    cylindrical shell 124

65. Specific inductive capacity 125

66. The charge induced on a conducting sphere by a charge
    at an outside point 130

67. The energy of charged conductors 134

THE

NEWTONIAN POTENTIAL FUNCTION.

———•o⊹⊛⊹o•———

CHAPTER I.

THE ATTRACTION OF GRAVITATION.

1. The Law of Gravitation. Every body in the universe attracts every other body with a force which depends for magnitude and direction upon the masses of the two bodies and upon their relative positions.

An *approximate* value of the attraction between any two rigid bodies may be obtained by imagining the bodies to be divided into small particles, and assuming that every particle of the one body attracts every particle of the other with a force directly proportional to the product of the masses of the two particles, and inversely proportional to the square of the distance between their centres or other corresponding points. The *true* value of the attraction is the limit approached by this approximate value as the particles into which the bodies are supposed to be divided are made smaller and smaller.

2. The Attraction at a Point. By "the attraction *at* any point *P* in space, due to one or more attracting masses," is meant the limit which would be approached by the value of the attraction on a sphere of unit mass centred at *P* if the radius of the sphere were made continually smaller and smaller while its mass remained unchanged. The attraction at *P* is, then, the attraction on a unit mass supposed to be *concentrated* at *P*.

If the attraction at every point throughout a certain region has a value other than zero, the region is called "a field of force"; and the attraction at any point P in the region is called "the strength of the field" at that point.

3. The Unit of Force. It will presently appear that all spheres made of homogeneous material attract bodies outside of themselves as if the masses of the spheres were concentrated at their middle points. If, then, k be the force of attraction between two unit masses concentrated at points at the unit distance apart, the attraction at a point P due to a homogeneous sphere of radius a and of density ρ is $k \cdot \dfrac{4 \pi a^3 \rho}{3 r^2}$, where r is the distance of P from the centre of the sphere. In all that follows, however, we shall take as our *unit of force* the force of attraction between two unit masses concentrated at points at the unit distance apart. Using these units, k in the expression given above becomes 1, and the attraction between two particles of mass m_1 and m_2 concentrated at points r units apart is $\dfrac{m_1 m_2}{r^2}$.

4. Attraction due to Discrete Particles. The attraction at a point P, due to particles concentrated at different points in the same plane with P, may be expressed in terms of two components at right angles to each other.

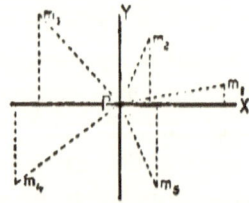

FIG. 1.

Let the straight lines joining P with the different particles be denoted by r_1, r_2, r_3, $\cdots$, and the angles which these lines make with some fixed line Px by a_1, a_2, a_3, $\cdots$. If, then, the masses of the several particles are respectively m_1, m_2, m_3, $\cdots$, the components of the attraction at P are

$$X = \frac{m_1 \cos a_1}{r_1^2} + \frac{m_2 \cos a_2}{r_2^2} + \cdots = \sum \frac{m \cos a}{r^2} \qquad [1]$$

in the direction Px, and

$$Y = \frac{m_1 \sin a_1}{r_1^2} + \frac{m_2 \sin a_2}{r_2^2} + \cdots = \sum \frac{m \sin a}{r^2} \qquad [2]$$

in the direction Py, perpendicular to Px.

The resultant force at P is

$$R = \sqrt{X^2 + Y^2}, \qquad [3]$$

and its line of action makes with Px the angle whose tangent is $\dfrac{Y}{X}$.

If the particles do not all lie in the same plane with P, we may draw through P three mutually perpendicular axes, and call the angles which the lines joining P with the different particles make with the first axis a_1, a_2, a_3, $\cdots$; with the second axis, β_1, β_2, β_3, $\cdots$; and with the third axis, γ_1, γ_2, γ_3, $\cdots$. The three components in the directions of these axes of the attraction at P due to all the particles are then

$$X = \sum \frac{m \cos a}{r^2}; \quad Y = \sum \frac{m \cos \beta}{r^2}; \quad Z = \sum \frac{m \cos \gamma}{r^2}. \qquad [4]$$

The resultant attraction is

$$R = \sqrt{X^2 + Y^2 + Z^2}, \qquad [5]$$

and its line of action makes with the axes angles whose cosines are respectively

$$\frac{X}{R}, \quad \frac{Y}{R}, \quad \text{and} \quad \frac{Z}{R}. \qquad [6]$$

5. Attraction at a Point in the Produced Axis of a Straight Wire. Let μ be the mass of the unit of length of a uniform straight wire AB of length l, and of cross section so small that

we may suppose the mass of the wire concentrated in its axis
(see Fig. 2), and let P be a point in the line AB produced at a

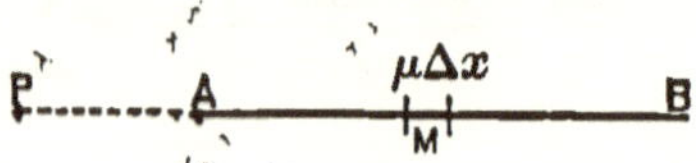

FIG. 2.

distance a from A. Divide the wire into elements of length
Δx. The attraction at P due to one of these elements, M, whose
nearest point is at a distance x from P, is less than $\dfrac{\mu \Delta x}{x^2}$ and
greater than $\dfrac{\mu \Delta x}{(x + \Delta x)^2}$.

The attraction at P due to the whole wire lies between
$\sum \dfrac{\mu \Delta x}{x^2}$ and $\sum \dfrac{\mu \Delta x}{(x + \Delta x)^2}$; but these quantities approach the
same limit as Δx is made to approach zero, so that the attrac-
tion at P is

$$\lim_{\Delta x \doteq 0} \sum \frac{\mu \Delta x}{x^2} = \int_a^{a+l} \frac{\mu dx}{x^2} = \mu \left[\frac{1}{a} - \frac{1}{a + l} \right]. \qquad [7]$$

If the coördinates of P, A, and B are respectively $(x, 0, 0)$,
$(x_1, 0, 0)$, and $(x_1 + l, 0, 0)$, this result may be put into the form

$$\mu \left[\frac{1}{x_1 - x} - \frac{1}{x_1 - x + l} \right]. \qquad [8]$$

6. Attraction at any Point, due to a Straight Wire. Let P
(Fig. 3) be any point in the perpendicular drawn to the straight
wire AB at A, and let $PA = c$, $AB = l$, $AM = x$, and the angle
$ABP = \delta$. Let MN be one of the equal elements of mass ($\mu \Delta x$)
into which the wire is divided, and call PM, r. The attraction
at P due to this element is approximately equal to $\dfrac{\mu \Delta x}{r^2}$, and
acts in some direction lying between PM and PN. This attrac-
tion can be resolved into two components whose approximate
values are $\dfrac{\mu \Delta x \cdot c}{(c^2 + x^2)^{\frac{3}{2}}}$ in the direction PA, and $\dfrac{\mu \Delta x \cdot x}{(c^2 + x^2)^{\frac{3}{2}}}$ in the

direction PL. The true values of the components in these directions of the attraction at P, due to the whole wire, are, then, respectively :

$$\int_0^l \frac{\mu c\, dx}{(c^2 + x^2)^{\frac{3}{2}}} = \frac{\mu}{c}\left[\frac{x}{\sqrt{c^2 + x^2}}\right]_0^l = \frac{\mu}{c}\cos\delta, \qquad [9]$$

and

$$\int_0^l \frac{\mu x\, dx}{(c^2 + x^2)^{\frac{3}{2}}} = \frac{-\mu}{c}\left[\frac{c}{\sqrt{c^2 + x^2}}\right]_0^l = \frac{\mu}{c}(1 - \sin\delta). \qquad [10]$$

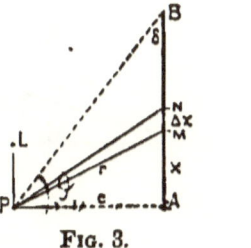

FIG. 3.

The resultant attraction is equal to the square root of the sum of the squares of these components, or

$$R = \frac{\mu}{c}\sqrt{2(1 - \sin\delta)} = \frac{\mu}{c}\sqrt{2(1 - \cos APB)} = \frac{2\mu}{c}\sin\tfrac{1}{2}APB. \quad [11]$$

and its line of action makes with PA an angle whose tangent is

$$\frac{1 - \sin\delta}{\cos\delta} = \frac{1 - \cos APB}{\sin APB} = \frac{2\sin^2\tfrac{1}{2}APB}{2\sin\tfrac{1}{2}APB\cdot\cos\tfrac{1}{2}APB} = \tan\tfrac{1}{2}APB.$$

That is, the resultant attraction at P acts in the direction of the bisector of the angle APB.

From these results we can easily obtain the value of the attraction at any point P, due to a uniform straight wire $B'B$ (Fig. 4). Drop a perpendicular PA from P upon the axis of the wire. Let $AB = l$, $AB' = l'$, $PA = c$, $ABP = \delta$, $AB'P = \delta'$, $BPB' = \theta$. The component in the direction PA of the attraction at P is [9]

$$\frac{\mu}{c}(\cos\delta + \cos\delta'),$$

and that in the direction PL is

$$\frac{\mu}{c}(\sin\delta' - \sin\delta),$$

so that the resultant attraction is

$$R = \frac{\mu}{c}\sqrt{2[1+\cos(\delta+\delta')]} = \frac{2\mu}{c}\cos\tfrac{1}{2}(\delta+\delta') = \frac{2\mu}{c}\sin\tfrac{1}{2}\theta. \quad [12]$$

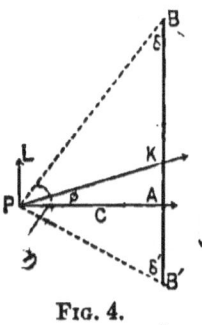

Fig. 4.

The line of action PK of R makes with PA an angle ϕ such that

$$\tan\phi = \frac{\sin\delta' - \sin\delta}{\cos\delta + \cos\delta'} = \tan\tfrac{1}{2}(\delta'-\delta); \quad [13]$$

$$\therefore B'PK = \frac{\pi}{2} - \delta' + \tfrac{1}{2}(\delta'-\delta) = \frac{\pi}{2} - \tfrac{1}{2}(\delta+\delta'),$$

and

$$BPK = \frac{\pi}{2} - \delta - \tfrac{1}{2}(\delta'-\delta) = \frac{\pi}{2} - \tfrac{1}{2}(\delta+\delta').$$

It is to be noticed that PK bisects the angle θ, and does not in general pass through the centre of gravity or any other fixed point of the wire. Indeed, the path of a particle moving from rest under the attraction of a straight wire is generally curved; for if the particle should start at a point Q and move a short distance on the bisector of the angle BQB' to Q', the attraction of the wire would now urge the particle in the direction of the bisector of the angle $BQ'B'$, and this is usually not coincident with the bisector of BQB'.

If q is the area of the cross section of the wire, and ρ the mass of the unit volume of the substance of which the wire is made, we may substitute for μ in the formulas of this section its value $q\rho$.

If instead of a very thin wire we had a body in the shape of a prism or cylinder of considerable cross section, we might divide this up into a large number of slender prisms and use the equations just obtained to find the limit of the sum of the attractions at any point due to all these elementary prisms. This would be the attraction due to the given body.

7. Attraction at a Point in the Produced Axis of a Cylinder of Revolution. In order to find the attraction due to a homogeneous cylinder of revolution at any point P (Fig. 5) in the axis of the cylinder produced, it will be convenient to imagine the cylinder cut up into discs of constant thickness Δc, by means of planes perpendicular to the axis.

Let ρ be the mass of the unit of volume of the cylinder, and a the radius of its base. Consider a disc whose nearer face is at a distance c from P, and divide it into elements by means of

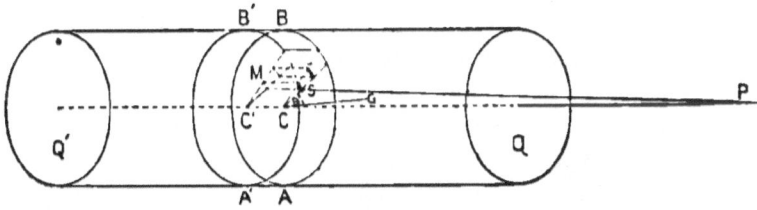

FIG. 5.

radial planes drawn at angular intervals of $\Delta\theta$ and concentric cylindrical surfaces at radial intervals of Δr.

The mass of any element M whose inner radius is r is equal to $\rho\Delta c\cdot\Delta\theta[r\Delta r+\frac{1}{2}(\Delta r)^2]$, and the whole attraction at P due to M is approximately $\rho\dfrac{\Delta\theta\Delta c[r\Delta r+\frac{1}{2}(\Delta r)^2]}{c^2+r^2}$ in a line joining P with some point of M. The component of this attraction in the direction PC is found by multiplying the expression just given by $\dfrac{c}{\sqrt{c^2+r^2}}$, the cosine of the angle CPS, so that the attraction at P in the direction PC, due to the whole disc, is approximately

$$\Delta c \cdot \lim \sum \frac{\rho c\, \Delta\theta\left[r\Delta r + \frac{1}{2}(\Delta r)^2\right]}{(c^2+r^2)^{\frac{3}{2}}} = \Delta c \int_0^{2\pi}\!\!d\theta \int_0^a \frac{\rho c r\, dr}{(c^2+r^2)^{\frac{3}{2}}}$$

$$= 2\pi\rho\,\Delta c\left[1 - \frac{c}{\sqrt{c^2+a^2}}\right]. \qquad [14]$$

If the bases of the cylinder are at distances c_0 and c_0+h from P, the true value of the attraction at P in the direction PC, due to the cylinder QQ', is

$$\lim_{\Delta c \doteq 0} \sum 2\pi\rho\,\Delta c\left[1 - \frac{c}{\sqrt{c^2+a^2}}\right] = 2\pi\rho \int_{c_0}^{c_0+h}\!\!\left(1 - \frac{c}{\sqrt{c^2+a^2}}\right) dc$$

$$= 2\pi\rho[h + \sqrt{c_0^2+a^2} - \sqrt{(c_0+h)^2+a^2}]. \qquad [15]$$

This is evidently the whole attraction at P due to the cylinder, for considerations of symmetry show us that the resultant attraction at P has no component perpendicular to PC.

[14] gives the attraction due to the elementary disc $ABA'B'$, on the assumption that the whole matter of the disc is concentrated at the face ABC. The actual attraction at P due to this disc may be found by putting $c_0 = c$ and $h = \Delta c$ in [15].

If a, the radius of the cylinder, is very large compared with h and c_0, the expression [15] for the attraction at P due to the cylinder approaches the value $2\pi\rho h$.

8. Attraction at the Vertex of a Cone. The attraction due to a homogeneous cone of revolution, at a point at the vertex of the cone, may be found by the aid of [14].

If Fig. 6 represents a plane section of the cone taken through the axis, and if $PM = c$, $MM' = \Delta c$, and $MB = r$, the attraction at P due to the disc $ABCD$ is approximately

$$2\pi\rho\,\Delta c\left[1 - \frac{c}{\sqrt{c^2+r^2}}\right] = 2\pi\rho\,\Delta c\,(1 - \cos a),$$

and the attraction due to the whole cone is

$$\underset{\Delta c \doteq 0}{\text{limit}} \sum 2\pi\rho(1 - \cos a)\Delta c = 2\pi\rho(1 - \cos a) \underset{\Delta c \doteq 0}{\text{limit}} \sum \Delta c$$

$$= 2\pi\rho(1 - \cos a) \cdot PL. \qquad [16]$$

. The attraction at P due to the frustum $ABKN$ is found by subtracting the value of the attraction due to the cone ABP from the expression given in [16]. The result is

$$2\pi\rho(1 - \cos a)(PL - PM) = 2\pi\rho(1 - \cos a)ML, \qquad [17]$$

and it is easy to see from this that discs of equal thickness cut out of a cone of revolution at different distances from the vertex by planes perpendicular to the axis exert equal attractions at the vertex of the cone.

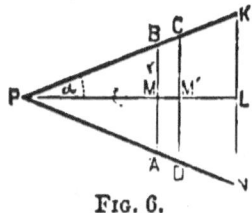

FIG. 6.

It follows almost directly that the portions cut out of two concentric spherical shells of equal uniform density and equal thickness, by *any* conical surface having its vertex at the common centre P of the shells, exert equal attraction at this centre; but we may prove this proposition otherwise, as follows:

Divide the inner surface of the portion cut out of one of the shells by the given cone into elements, and make the perimeter of each of these surface elements the directrix of a conical surface having its vertex at P. Divide the given shells into elementary shells of thickness Δr by means of concentric spherical surfaces drawn about P. In this way the attracting masses will be cut up into volume elements.

Let ML' (Fig. 7) represent one of these elements, whose inner surface has a radius equal to r; then, if the elementary cone APB intercept an element of area $\Delta\omega$ from a spherical surface of radius unity drawn around P, the area of the surface element at MM' is $r^2\Delta\omega$, and that at LL' is $(r+\Delta r)^2\Delta\omega$. The

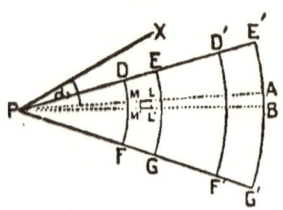

Fig. 7.

attraction at P in the direction PM, due to the element ML', is approximately

$$\rho\frac{r^2\Delta\omega\Delta r}{r^2}=\rho\Delta\omega\Delta r,$$

and the component of this in any direction Px, making an angle a with PM, is approximately $\rho\Delta\omega\Delta r\cos a$. The attraction at P in the direction Px, due to the whole shell $EDFG$, is, then,

$$X=\lim\sum\rho\Delta r\Delta\omega\cos a,$$

where the sum is to include all the volume elements which go to make up the shell. If $PF=r_0$, $PG=r_1$, $PF'=r_0'$, $PG'=r_1'$, and $\mu=FG=F'G'$,

$$X=\int_{r_0}^{r_1}dr\int\cos a\,d\omega=\rho\mu\int\cos a\,d\omega.$$

The attraction at P in the same direction, due to the shell $E'D'F'G'$, is

$$X'=\rho\int_{r_0'}^{r_1'}dr\int\cos a\,d\omega=\rho\mu\int\cos a\,d\omega.$$

But the limits of integration with regard to ω are the same in both cases; $\therefore X=X'$, which was to be proved.

If the shells are of different thicknesses, it is evident that they will exert attractions at P proportional to these thicknesses.

The area of the portion which a conical surface cuts out of a spherical surface of unit radius drawn about the vertex of the cone is called "the solid angle" of the conical surface.

9. Attraction of a Spherical Shell. In order to find the attraction at P, any point in space, due to a homogeneous spherical shell of radii r_0 and r_1, it will be best to begin by dividing up the shell into a large number of concentric shells of thickness Δr, and to consider first the attraction of one of these thin shells, whose inside radius shall be r.

Let ρ be the density of the given shell, that is, the mass of the unit of volume of the material of which the shell is composed. Join P (Fig. 8) with O by a straight line cutting the inner surface of the thin shell at N, and pass a plane through PO cutting this inner surface in a great circle $NLSL'$, which

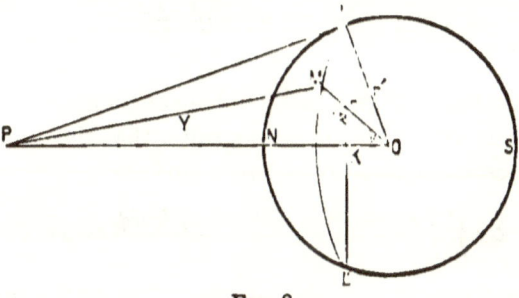

FIG. 8.

will serve as a prime meridian. Using N as a pole, describe upon the inner surface of the thin shell a number of parallels of latitude so as to cut off equal arcs on $NLSL'$. Denote by $\Delta\theta$ the angle which each one of these arcs subtends at O. Through PO pass a number of planes so as to cut up each parallel of latitude into equal arcs. Denote by $\Delta\phi$ the angle between any two contiguous planes of this series. By this means the inner surface of the elementary shell will be divided into small quadrilaterals, each of which will have two sides formed of meridian arcs, of length $r \cdot \Delta\theta$, and two sides formed of arcs of parallels of latitude, of length $r \sin\theta \cdot \Delta\phi$ and $r \sin(\theta + \Delta\theta) \cdot \Delta\phi$, where

θ is the angle which the radius drawn to the parallel of higher latitude makes with ON. The area of one of these quadrilaterals is approximately $r^2 \sin\theta \cdot \Delta\theta \cdot \Delta\phi$, and the thickness of the shell is Δr, so that the element of volume is approximately $r^2 \sin\theta \cdot \Delta r \cdot \Delta\theta \cdot \Delta\phi$. Let $PM = y$, then the attraction at P, due to an element of mass which has a corner at M, is approximately $\dfrac{\rho r^2 \sin\theta \Delta r \Delta\theta \Delta\phi}{y^2}$, in the direction PM.

This force may be resolved into three components: one in the direction PO, the others in directions perpendicular to PO and to each other; but it is evident from considerations of symmetry that in finding the attraction at P due to the whole shell we shall need only that component which acts in PO. This is approximately $\dfrac{\rho r^2 \sin\theta \cdot \Delta r \Delta\theta \Delta\phi \cdot \cos KPM}{y^2}$; or, if $PO = c$,

$$\frac{\rho r^2 \sin\theta (c - r\cos\theta)\Delta r \Delta\theta \Delta\phi}{y^3}. \qquad [18]$$

The attraction at P due to the whole elementary shell is, then, approximately (truly on the assumption that the whole mass of the shell is concentrated at its inner surface),

$$\Delta r \int\int \frac{\rho r^2 \sin\theta \, (c - r\cos\theta) d\theta \, d\phi}{y^3} = \Delta r X; \qquad [19]$$

and the true value at P of the attraction due to the given shell is

$$\int_{r_0}^{r_1} X dr. \qquad [20]$$

If in the expression for X we substitute for θ its value in terms of y, we have, since

$$y^2 = c^2 + r^2 - 2cr\cos\theta,$$

and hence $2y \, dy = 2cr \sin\theta \, d\theta,$

$$X = \int_0^{2\pi} d\phi \int_{y_0}^{y_1} \frac{\rho r \, dy}{2c^2 y^2}(c^2 - r^2 + y^2) = \frac{\pi\rho r}{c^2} \int_{y_0}^{y_1} \left(\frac{c^2 - r^2}{y^2} + 1\right) dy$$

$$= \frac{\pi\rho r}{c^2}\left[\frac{r^2 - c^2 + y^2}{y}\right]_{y_0}^{y_1}. \qquad [21]$$

In order to find the limits of the integration with regard to y, we must distinguish between two cases:

I. If P is a point in the cavity enclosed by the given shell,

$$y_0 = r - c \quad \text{and} \quad y_1 = r + c;$$

$$\therefore X = \frac{\pi \rho r}{c^2}\left[\frac{r^2 - c^2 + (r+c)^2}{r+c} - \frac{r^2 - c^2 + (r-c)^2}{r-c}\right] = 0, \quad [22]$$

and
$$\int_{r_0}^{r_1} X dr = 0; \quad [23]$$

so that a homogeneous spherical shell exerts no attraction at points in the cavity which it encloses.

II. If P is a point without the given shell,

$$y_0 = c - r \quad \text{and} \quad y_1 = c + r;$$

$$\therefore X = \frac{\pi \rho r}{c^2}\left[\frac{r^2 - c^2 + (c+r)^2}{c+r} - \frac{r^2 - c^2 + (c-r)^2}{c-r}\right] = \frac{4\pi\rho r^2}{c^2}, \quad [24]$$

and
$$\int_{r_0}^{r_1} X dr = \frac{4}{3}\frac{\pi\rho}{c^2}(r_1^3 - r_0^3). \quad [25]$$

From this it follows that the attraction due to a spherical shell of uniform density is the same, at a point without the shell, as the attraction due to a mass equal to that of the shell concentrated at the shell's centre.

If in [25] we make $r_0 = 0$, we have the attraction, due to a solid sphere of radius r_1 and density ρ, at a point outside the sphere at a distance c from the centre. This is

$$\frac{4\pi\rho r_1^3}{3c^2}. \quad [26]$$

10. Attraction due to a Hemisphere. At any point P in the plane of the base of a homogeneous hemisphere, the attraction of the hemisphere gives rise to two components, one directed toward the centre of the base, the other perpendicular to the plane of the base. We will compute the values of these components for the particular case where P lies on the rim of the hemisphere's base, and for this purpose we will take the origin of our system of polar coördinates at P, because by so doing we shall escape having to deal with a quantity which becomes infinite at one of the limits of integration. Denote the coördinates of any point L in the hemisphere by r, θ, ϕ, where (Fig. 9) $XPN = \phi$, $IPL = \theta$, and $PL = r$.

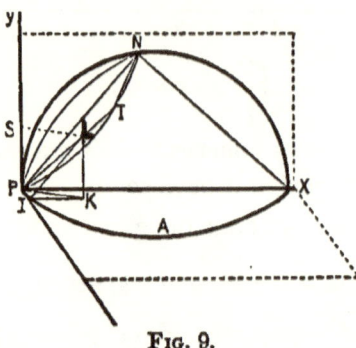

FIG. 9.

If r_1 be the radius of the hemisphere,

$$PT = PN \cos NPT = PX \cos XPN \cdot \cos NPT = 2\,r_1 \sin \theta \cos \phi.$$

$$\cos XPL = \frac{IK}{PL} = \frac{IK}{r} = \frac{IL \cos \phi}{r} = \sin \theta \cos \phi.$$

$$\cos SPL = \frac{PS}{PL} = \frac{KL}{r} = \frac{IL \sin \phi}{r} = \sin \theta \sin \phi.$$

The mass of a polar element of volume whose corner is at L is approximately $\rho \cdot IL \Delta \phi \cdot PL \Delta \theta \cdot \Delta r$ or $\rho r^2 \sin \theta \Delta r \Delta \theta \Delta \phi$, and this divided by r^2 is the attraction at P in the direction PL of the element, supposed concentrated at L. The components of this attraction in the direction PX and PY are respectively $\rho \sin \theta \Delta r \Delta \theta \Delta \phi \cos XPL$ and $\rho \sin \theta \Delta r \Delta \theta \Delta \phi \cos SPL$.

The component in the direction Py of the attraction at P due to the whole hemisphere is, then,

$$\int_0^{\frac{\pi}{2}} d\phi \int_0^{\pi} d\theta \int_0^{2\,r_1 \sin \theta \cos \phi} \rho \sin^2 \theta \sin \phi \, dr = \tfrac{4}{3} \rho r_1, \qquad [27]$$

and the component in the direction Px is

$$\int_0^{\frac{\pi}{2}} d\phi \int_0^{\pi} d\theta \int_0^{2r_1 \sin\theta \cos\phi} \rho \sin^2\theta \cos\phi \, dr = \tfrac{2}{3}\pi\rho r_1. \qquad [28]$$

This last expression might have been obtained from [26] by making c equal to r and halving the result.

11. Attraction of a Hemispherical Hill. If at a point on the earth at the southern extremity of a homogeneous hemispherical hill of density ρ and radius r_1 the force of gravity due to the earth, supposed spherical, is g, the attraction due to the earth and the hill will give rise to two components, $g - \tfrac{4}{3}\rho r_1$ downwards, and $\tfrac{2}{3}\pi\rho r_1$ northwards. The resultant attraction does not therefore act in the direction of the centre of the earth, but makes with this direction an angle whose tangent is $\dfrac{\tfrac{2}{3}\pi\rho r_1}{g - \tfrac{4}{3}\rho r_1}$.

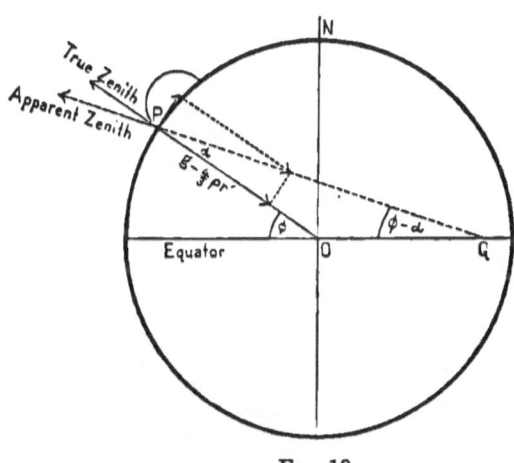

FIG. 10.

Let ϕ (Fig. 10) be the true latitude of the place and $(\phi - \alpha)$ the apparent latitude, as obtained by measuring the angle which the plumb-line at the place makes with the plane of the equator. Let a be the radius of the earth and σ its average density. Then

$$\tan \alpha = \frac{\tfrac{2}{3}\pi\rho r_1}{g - \tfrac{4}{3}\rho r_1} = \frac{\pi\rho r_1}{2(\pi a\sigma - \rho r_1)}. \qquad [29]$$

The radius of the earth is very large compared with the radius of the hill, and a is a small angle, so that approximately $a = \frac{\rho''_1}{2\,a\sigma}$, and the apparent latitude of the place is $\phi - \frac{\rho''_1}{2\,a\sigma}$.

If ϕ_1 is the true latitude of a place just north of the same hill, its apparent latitude will be $\phi_1 + \frac{\rho''_1}{2\,a\sigma}$, and the apparent difference of latitude between the two places, one just north of the hill and the other just south of it, will be the true difference plus $\frac{\rho''_1}{a\sigma}$.   If there were a hemispherical cavity between the two places instead of a hemispherical hill, the apparent difference of latitude would be less than the true difference.

12. Ellipsoidal Homœoids.   A shell, thick or thin, bounded by two ellipsoidal surfaces, concentric, similar, and similarly placed, shall be called an *ellipsoidal homœoid*.

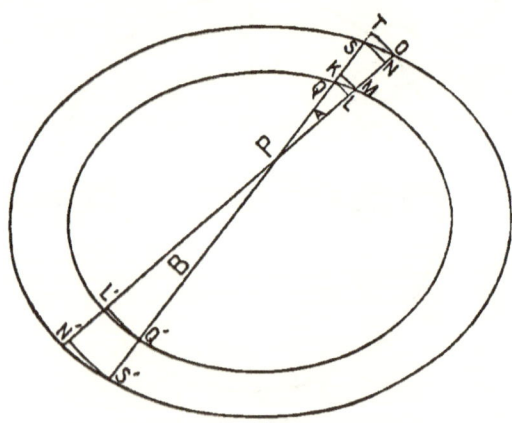

Fig. 11.

It is a property of every such shell that if any straight line cut its outer surface at the points S, S' (Fig. 11) and its inner surface at Q, Q', so that these four points lie in the order $SQQ'S'$, the length SQ will be equal to the length $Q'S'$.

We will prove that the attraction of a homogeneous closed

ellipsoidal homœoid, at any point P in the cavity which it shuts in, is zero.

Make P the vertex of a slender conical surface of two nappes, A and B, and suppose the plane of the paper to be so chosen that PQ is the shortest and PM the longest length cut from any element of the nappe A by the inner surface of the homœoid. Draw about P spherical surfaces of radii PQ, PM, PS, and PO, and imagine the space between the innermost and outermost of these surfaces filled with matter of the same density as the homœoid. The nappe A cuts out a portion from this spherical shell whose trace on the plane of the paper is $QLOT$. Let us call this, for short, "the element $QLOT$." The attraction at P, due to the element $QMOS$ which A cuts out of the homœoid, is less than the attraction at the same point due to the element $QLOT$, and greater than that due to the element whose trace is $KMNS$. But the attraction at P, due to the first of these elements of spherical shells, is to the attraction due to the other as the thickness of the first shell is to that of the other, or as QT is to KS. (See Section 8.) The limit of the ratio of QT to KS, as the solid angle of the cone is made smaller and smaller, is unity; therefore the limit of the ratio of the attraction at P due to the element $QMOS$, to the attraction due to the element of spherical shell whose trace is $QLNS$, is unity. By a similar construction it is easy to show that the limit of the ratio of the attraction at P, due to the element which B cuts out of the homœoid, to the attraction due to the portion of spherical shell whose trace is $Q'L'N'S'$, is unity.

But the attractions at P, due to the elements $Q'L'N'S'$ and $QLNS$, are equal in amount (since their thicknesses are the same) and opposite in direction, so that if for the elements of the homœoid these elements were substituted, there would be no resultant attraction at P. In order to get the attraction at P in any direction due to the whole homœoid we may cut up the inner surface of the homœoid into elements, use the perimeter of each one of these elements as the directrix of a conical surface having its vertex at P, and find the limit of the sum of the attractions due to the elements which these conical surfaces cut from the homœoid. Wherever we have to find the finite limit of the sum of a series of infinitesimal quantities, we may without error substitute for any one of these another infinitesimal, the limit of whose ratio to the first is unity. For the attractions at P due to the elements of the homœoid we may, therefore, substitute attractions due to elements of spherical shells, which, as we have seen, destroy each other in pairs. Hence our proposition.

A shell bounded by two concentric spherical surfaces gives a special case under this theorem.

13. Sphere of Variable Density. The density of a homogeneous body is the amount of matter contained in the unit volume of the material of which the body is composed, and this may be obtained by dividing the mass of the body by its volume.

If the amount of matter contained in a given volume is not the same throughout a body, the body is called heterogeneous, and its density is said to be variable.

The average density of a heterogeneous body is the ratio of the mass of the body to its volume. The actual density ρ at any point Q inside the body is defined to be the limit of the ratio of the mass of a small portion of the body taken about Q to the volume of this portion as the latter is made smaller and smaller.

The attraction, at any point P, due to a spherical shell whose density is the same at all points equidistant from the common centre of the spherical surfaces which bound the shell but different at different distances from this centre, may be obtained with the help of some of the equations in Article 9.

Since ρ is independent of θ and ϕ, it may be taken out from under the signs of integration with regard to these variables, although it must be left under the sign of integration with regard to r.

Equations 19 to 24 inclusive hold for the case that we are now considering as well as for the case when ρ is constant, so that the attraction at all points within the cavity enclosed by a spherical shell whose density varies with the distance from the centre is zero.

If P is without the shell, the attraction is

$$\int_{r_0}^{r_1} \mathrm{X} dr = \int_{r_0}^{r_1} \frac{4\pi\rho r^2 dr}{c^2},$$

or, if $\rho = f(r)$,

$$\frac{4\pi}{c^2} \int_{r_0}^{r_1} f(r) \cdot r^2 dr. \qquad [30]$$

The mass of the shell is evidently

$$\lim_{\Delta r \doteq 0} \sum_{r_0}^{r_1} 4\pi r^2 \cdot f(r) dr = 4\pi \int_{r_0}^{r_1} f(r) \cdot r^2 dr, \qquad [31]$$

and [30] declares that a spherical shell whose density is a function of the distance from its centre attracts at all outside points as if the whole mass of the shell were concentrated at the centre.

If $r_0 = 0$, we have the case of a solid sphere.

14. Attraction due to any Mass. In order to find the attraction at a point P (Fig. 12), due to any attracting masses M', we may choose a system of rectangular coördinate axes and divide

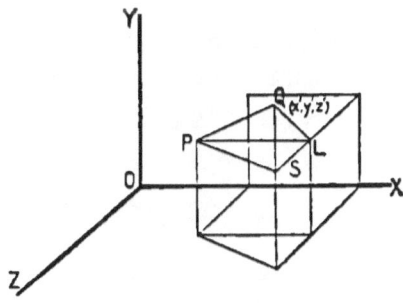

Fig. 12.

M' up into volume elements. If ρ is the average density of one of these elements $(\Delta v')$, the mass of the element will be $\rho \Delta v'$. Let Q, whose coördinates are x', y', z', be a point of the element, and let the coördinates of P be x, y, z. The attraction at P in the direction PQ due to this element is approximately $\frac{\rho \Delta v'}{\overline{PQ}^2}$, and the components of this in the direction of the coördinate axes are

$$\frac{\rho \Delta v'}{\overline{PQ}^2} \cos a', \quad \frac{\rho \Delta v'}{\overline{PQ}^2} \cos \beta', \quad \text{and} \quad \frac{\rho \Delta v'}{\overline{PQ}^2} \cos \gamma', \qquad [32]$$

where a', β', γ' are the angles which PQ makes with the positive directions of the axes.

It is easy to see that

$$\cos a' = \frac{PL}{PQ} = \frac{x' - x}{PQ},$$

and, similarly, that

$$\cos \beta' = \frac{y' - y}{PQ}, \quad \text{and} \quad \cos \gamma' = \frac{z' - z}{PQ}.$$

Moreover,

$$\overline{PQ}^2 = \overline{PL}^2 + \overline{LS}^2 + \overline{SQ}^2 = (x' - x)^2 + (y' - y)^2 + (z' - z)^2,$$

and this we will call r^2.

The true values of the components in the direction of the coördinate axes of the attraction at P, due to all the elements which go to make up M', are, then,

$$X = \lim_{\Delta v' \doteq 0} \sum \frac{\rho \Delta v'(x' - x)}{r^3}$$

$$= \iiint \frac{\rho(x' - x)\, dx'dy'dz'}{[(x' - x)^2 + (y' - y)^2 + (z' - z)^2]^{\frac{3}{2}}}; \qquad [33_A]$$

$$Y = \lim_{\Delta v' \doteq 0} \sum \frac{\rho \Delta v'(y' - y)}{r^3}$$

$$= \iiint \frac{\rho(y' - y)\, dx'dy'dz'}{[(x' - x)^2 + (y' - y)^2 + (z' - z)^2]^{\frac{3}{2}}}, \qquad [33_B]$$

$$Z = \lim_{\Delta v' \doteq 0} \sum \frac{\rho \Delta v'(z' - z)}{r^3}$$

$$= \iiint \frac{\rho(z' - z)\, dx'dy'dz'}{[(x' - x)^2 + (y' - y)^2 + (z' - z)^2]^{\frac{3}{2}}}; \qquad [33_C]$$

where ρ is the density at the point (x', y', z'), and where the integrations with regard to x', y', and z' are to include the whole of M'.

The resultant attraction at P, due to M', is

$$R = \sqrt{X^2 + Y'^2 + Z^2}; \qquad [34]$$

and its line of action makes with the coördinate axes angles whose cosines are

$$\lambda = \frac{X}{R}, \quad \mu = \frac{Y}{R}, \quad \text{and} \quad \nu = \frac{Z}{R}. \qquad [35]$$

The component of the attraction at the point (x, y, z) in a direction making an angle ϵ with the line of action of R is $R \cos \epsilon$. If the direction cosines of this direction are λ', μ', ν', we have

$$\cos \epsilon = \lambda \lambda' + \mu \mu' + \nu \nu'.$$

15. The quantities X, Y, Z, and R, which occur in the last section, are in general functions of the coördinates x, y, and z of the point P. Let us consider X, whose value is given in $[33_x]$. If P lies without the attracting mass M', the quantity $\frac{x' - x}{r^3}$ is finite for all the elements into which M' is divided. Let L be the largest value which it can have for any one of these elements, then X is less than $L \int \int \int \rho \, dx' dy' dz'$, or $L \cdot M'$, and this is finite. If P is a point within the space which the attracting mass occupies, it is easy to show that, whatever physical meaning we may attach to X, it has a finite value. To prove this, make P the origin of a system of polar coördinates, and divide M' up into elements like those used in Section 10. It will then be clear that

$$X = \int \int \int \rho \sin^2 \theta \cos \phi \, dr \, d\theta \, d\phi, \qquad [36]$$

where the limits are to be chosen so as to include all the attracting mass. Since $\sin^2 \theta \cos \phi$ can never be greater than unity, X is less than $\int\int\int \rho\, dr\, d\theta\, d\phi$, which is evidently finite when ρ is finite, as it always is in fact.

The corresponding expressions,

$$Y = \int\int\int \rho \sin^2\theta \sin\phi\, dr\, d\theta\, d\phi, \qquad [37]$$

and

$$Z = \int\int\int \rho \sin\theta \cos\theta\, dr\, d\theta\, d\phi, \qquad [38]$$

can be proved finite in a similar manner; and it follows that X, Y, Z, and consequently R, are finite for all values of x, y, and z.

As a special case, the attraction at a point P within the mass of a homogeneous spherical shell, of radii r_0 and r_1, and of density ρ, is

$$\tfrac{4}{3}\pi\rho\left(\frac{r^3 - r_0^3}{r^2}\right), \qquad [39]$$

where r is the distance of P from the centre of the shell.

16. Attraction between Two Straight Wires. Let AK and BK' (Fig. 13) be two straight wires of lengths l and l' and of line-densities μ and μ'; and let $KB = c$. Divide AK into

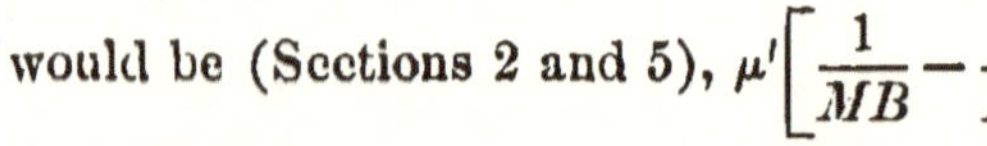

Fig. 13.

elements of length Δx, and consider one of these MM', such that $AM = x$. The attraction of BK' on a unit mass concentrated at M would be (Sections 2 and 5), $\mu'\left[\dfrac{1}{MB} - \dfrac{1}{MK'}\right]$. If, therefore, the whole element MM' whose mass is $\mu\Delta x$ were concentrated at M, the attraction on it, due to BK', would be

$$\mu\mu'\Delta x\left[\frac{1}{MB} - \frac{1}{MK'}\right] = \mu\mu'\Delta x\left[\frac{1}{l + c - x} - \frac{1}{l + l' + c - x}\right]. \qquad [40]$$

The actual force, due to the attraction of BK', with which the whole wire AK is urged toward the right, is

$$\underset{\Delta x \doteq 0}{\text{limit}} \sum\nolimits_0^{l} \mu\mu' \Delta x \left[\frac{1}{l+c-x} - \frac{1}{l+l'+c-x} \right]$$

$$= \mu\mu' \int_0^l \left(\frac{1}{x-(l+l'+c)} - \frac{1}{x-(l+c)} \right) dx$$

$$= \mu\mu' \left[\log \frac{x-l-l'-c}{x-l-c} \right]_0^l = \mu\mu' \log \frac{(l+c)(l'+c)}{c(l+l'+c)}. \quad [41]$$

17. Attraction between Two Spheres. Consider two homogeneous spheres of masses M and M' (Fig. 14), whose centres C and C' are at a distance c from each other. Divide the sphere M' into elements in the manner described in Section 9. The attraction due to M at any point P' outside of this sphere is, as we have seen, $\dfrac{M}{CP'^2}$, and its line of action is in the direction $P'C$.

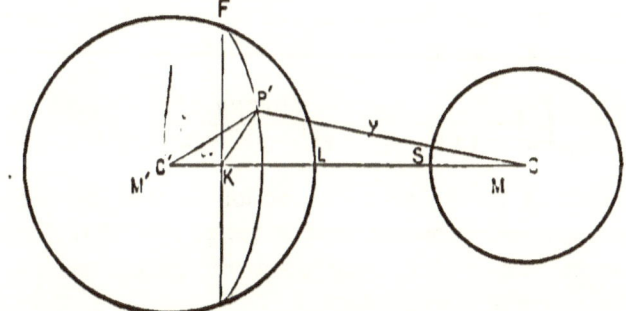

FIG. 14.

Let $P' = (r, \theta, \phi)$ be any point in the sphere M', and let $CP' = y$. The attraction of M in the direction $P'C$ on an element of mass $\rho r^2 \sin\theta \Delta r \Delta\theta \Delta\phi$ supposed concentrated at P' is $\dfrac{M\rho r^2 \sin\theta \, \Delta r \Delta\theta \Delta\phi}{y^2}$, and the component of this parallel to the line $C'C$ is $\dfrac{M\rho r^2 \sin\theta (c - r \cos\theta) \Delta r \Delta\theta \Delta\phi}{y^3}$. The force with which the whole sphere M' is urged toward the right by the attraction of M is, then,

$$M \int \int \int \frac{\rho r^2 \sin\theta \, dr \, d\theta \, d\phi (c - r\cos\theta)}{y^3}, \qquad [42]$$

where the integration is to be extended to all the elements which go to make up M'. It is proved in Section 9 that the value of this triple integral is $\frac{M'}{c^2}$, so that the force of attraction between the two spheres is $\frac{MM'}{c^2}$.

18. Attraction between any Two Rigid Bodies. In order to find the force with which a rigid body M is pulled in any direction (as for instance in that of the axis of x) by the attraction of another body M', we must in general find the value of a sextuple integral.

Let M be divided up into small portions, and let Δm be the mass of one of these elements which contains the point (x, y, z).

The component in the direction of the axis of x of the attraction at (x, y, z) due to M' is

$$\int \int \int \frac{\rho(x' - x) \, dx' dy' dz'}{[(x' - x)^2 + (y' - y)^2 + (z' - z)^2]^{\frac{3}{2}}},$$

and this would be the actual attraction in this direction on a unit mass supposed concentrated at (x, y, z). If the mass Δm were concentrated at this point, the attraction on it in the direction of the axis of x would be

$$\Delta m \int \int \int \frac{\rho(x' - x) \, dx' dy' dz'}{[(x' - x)^2 + (y' - y)^2 + (z' - z)^2]^{\frac{3}{2}}}. \qquad [43]$$

The actual attraction in the direction of the axis of x of M' upon the whole of M is, then,

$$\lim_{\Delta m \doteq 0} \sum \Delta m \cdot \int \int \int \frac{\rho(x' - x) \, dx' dy' dz'}{[(x' - x)^2 + (y' - y)^2 + (z' - z)^2]^{\frac{3}{2}}}. \qquad [44]$$

If ρ' is the density at the point (x, y, z), and if the elements into which M is divided are rectangular parallelopipeds of dimensions Δx, Δy, and Δz, the expression just given may be written

$$\int\int\int\int\int\int \frac{\rho'\rho(x'-x)\,dx\,dy\,dz\,dx'\,dy'\,dz'}{[(x'-x)^2+(y'-y)^2+(z'-z)^2]^{\frac{3}{2}}} \qquad [45]$$

where the integrations are first to be extended over M' and then over M.

EXAMPLES.

1. Find the resultant attraction, at the origin of a system of rectangular coördinates, due to masses of 12, 16, and 20 units respectively, concentrated at the points $(3, 4)$, $(-5, 12)$, and $(8, -6)$. What is its line of action?

2. Find the value, at the origin of a system of rectangular coördinates, of the attraction due to three equal spheres, each of mass m, whose centres are at the points $(a, 0, 0)$, $(0, b, 0)$, $(0, 0, c)$. Find also the direction-cosines of the line of action of this resultant attraction.

3. Show that the attraction, due to a uniform wire bent into the form of the arc of a circumference, is the same at the centre of the circumference as the attraction due to any uniform straight wire of the same density which is tangent to the given wire, and is terminated by the bounding radii (when produced) of the given wire.

4. Show that in the case of an oblique cone whose base is any plane figure the attraction at the vertex of the cone due to any frustum varies, other things being equal, as the thickness of the frustum.

5. Find the equation of a family of surfaces over each one of which the resultant force of attraction due to a uniform straight wire is constant.

6. Using the foot-pound-second system of fundamental units, and assuming that the average density of the earth is 5.6, compare with the poundal the unit of force used in this chapter.

7. If in Fig. 2 we suppose P moved up to A, the attraction at P becomes infinite according to [7], and yet Section 15 asserts that the value, at any point inside a given mass, of the attraction due to this mass is always finite. Explain this.

8. A spherical cavity whose radius is r is made in a uniform sphere of radius $2r$ and mass m in such a way that the centre of the sphere lies on the wall of the cavity. Find the attraction due to the resulting solid at different points on the line joining the centre of the sphere with the centre of the cavity.

9. A uniform sphere of mass m is divided into halves by the plane AB passed through its centre C. Find the value of the attraction due to each of these hemispheres at P, a point on the perpendicular erected to AB at C, if $CP = a$.

10. Considering the earth a sphere whose density varies only with the distance from the centre, what may we infer about the law of change of this density if a pendulum swing with the same period on the surface of the earth and at the bottom of a deep mine? What if the force of attraction increases with the depth at the rate of $\frac{1}{n}$th of a dyne per centimetre of descent?

11. The attraction due to a cylindrical tube of length h and of radii R_0 and R_1, at a point in the axis, at a distance c_0 from the plane of the nearer end, is

$$2\pi\rho\left[\sqrt{c_0^2+R_1^2}-\sqrt{c_0^2+R_0^2}+\sqrt{(c_0+h)^2+R_0^2}-\sqrt{(c_0+h)^2+R_1^2}\right].$$
[Stone.]

12. A spherical cavity of radius b is hollowed out in a sphere of radius a and density ρ, and then completely filled with matter, of density ρ_0. If c is the distance between the centre of the cavity and the centre of the sphere, the attraction due to the composite solid at a point in the line joining these two centres, at a distance d from the centre of the sphere, is

$$\frac{4}{3}\pi\left[\frac{\rho a^3}{d^2}+\frac{b^3(\rho_0-\rho)}{(d\pm c)^2}\right]. \qquad \text{[Stone.]}$$

13. The centre of a sphere of aluminum of radius 10 and of density 2.5, is at the distance 100 from a sphere of the same

size made of gold, of density 19. Show that the attraction due to these spheres is nothing at a point between them, at a distance of about 26.6 from the centre of the aluminum sphere.

[Stone.]

14. Show that the attraction at the centre of a sphere of radius r, from which a piece has been cut by a cone of revolution whose vertex is at the centre, is $\pi \rho r \sin^2 a$, where a is the half angle of the cone. [Stone.]

15. An iron sphere of radius 10 and density 7 has an eccentric spherical cavity of radius 6, whose centre is at a distance 3 from the centre of the sphere. Find the attraction due to this solid at a point 25 units from the centre of the sphere, and so situated that the line joining it with this centre makes an angle of 45° with the line joining the centre of the sphere and the centre of the cavity. [Stone.]

16. If the piece of a spherical shell of radii r_0 and r_1, intercepted by a cone of revolution whose solid angle is ω and whose vertex is the centre of the shell, be cut out and removed, find the attraction of the remainder of the shell at a point P situated in the axis of the cone at a given distance from the centre of the sphere. If in the vertical shaft of a mine a pendulum be swung, is there any appreciable error in assuming that the only matter whose attraction influences the pendulum lies nearer the centre of the earth, supposed spherical, than the pendulum does?

17. Show that the attraction of a spherical segment is, at its vertex,

$$2\pi h \rho \left\{ 1 - \frac{1}{3}\sqrt{\frac{2h}{a}} \right\},$$

where a is the radius of the sphere and h the height of the segment.

18. Show that the resultant attraction of a spherical segment on a particle at the centre of its base is

$$\frac{2\pi h \rho}{3(a-h)^2} \left[3a^2 - 3ah + h^2 - (2a-h)^{\frac{3}{2}} h^{\frac{1}{2}} \right].$$

19. Show that the attraction at the focus of a segment of a paraboloid of revolution bounded by a plane perpendicular to the axis at a distance b from the vertex is of the form

$$4\pi\rho a \log \frac{a+b}{a}.$$

20. Show that the attraction of the oblate spheroid formed by the revolution of the ellipse of semiaxes a, b, and eccentricity e, is, at the pole of the spheroid,

$$\frac{4\pi\rho b}{e^2}\left\{1 - \frac{(1-e^2)^{\frac{1}{2}}}{e}\sin^{-1}e\right\},$$

and that the attraction due to the corresponding prolate spheroid is, at its pole,

$$\frac{4\pi\rho a(1-e^2)}{e^2}\left\{\frac{1}{2e}\log\frac{1+e}{1-e} - 1\right\}.$$

21. Show that the attraction at the point $(c, 0, 0)$, due to the homogeneous solid bounded by the planes $x = a$, $x = b$, and by the surface generated by the revolution about the axis of x of the curve $y = f(x)$, is

$$2\pi\rho\int_a^b \left\{1 - \frac{c-x}{[(c-x)^2 + (fx)^2]^{\frac{1}{2}}}\right\}dx.$$

22. Prove that the attraction of a uniform lamina in the form of a rectangle, at a point P in the straight line drawn through the centre of the lamina at right angles to its plane, is

$$4\mu \sin^{-1}\frac{ab}{\sqrt{a^2+c^2}\sqrt{b^2+c^2}},$$

where $2a$ and $2b$ are the dimensions of the lamina and c the distance of P from its plane. [See Todhunter's *Analytical Statics*.]

CHAPTER II.

THE NEWTONIAN POTENTIAL FUNCTION IN THE CASE OF GRAVITATION.

19. Definition. If we imagine an attracting body M to be cut up into small elements, and add together all the fractions formed by dividing the mass of each element by the distance of one of its points from a given point P in space, the limit of this sum, as the elements are made smaller and smaller, is called the value at P of "the potential function due to M."

If we call this quantity V, we have

$$V = \lim_{\Delta m \doteq 0} \sum \frac{\Delta m}{r}, \qquad [46]$$

where Δm is the mass of one of the elements and r its distance from P, and where the summation is to include all the elements which go to make up M.

If we denote by ρ the average density of the element whose mass is Δm, and call the coördinates of the corner of this element nearest the origin x', y', z', and those of P, x, y, z, we may write

$$\Delta m = \bar{\rho} \Delta x' \Delta y' \Delta z',$$

and

$$V = \iiint \frac{\rho\, dx'dy'dz'}{\left[(x'-x)^2 + (y'-y)^2 + (z'-z)^2 \right]^{\frac{1}{2}}}, \qquad [47]$$

where ρ is the density at the point (x', y', z'), and where the triple integration is to include the whole of the attracting mass M.

As the position of the point P changes, the value of the quantity under the integral signs in [47] changes, and in general V is a function of the three space coördinates, i.e., $V = f(x, y, z)$.

To avoid circumlocution, a point at which the value of the

potential function is V_0 is sometimes said to be "at potential V_0." From the definition of V it is evident that if the value at a point P of the potential function due to a system of masses M_1 existing alone is V_1, and if the value at the same point of the potential function due to another system of masses M_2 existing alone is V_2, the value at P of the potential function due to M_1 and M_2 existing together is $V = V_1 + V_2$.

20. The Derivatives of the Potential Function. If P is a point outside the attracting mass, the quantity

$$\sqrt{(x'-x)^2 + (y'-y)^2 + (z'-z)^2},$$

which enters into the expression for V in [47], can never be zero, and the quantity under the integral signs is finite everywhere within the limits of integration; now, since these limits depend only upon the shape and position of the attracting mass and have nothing to do with the coördinates of P, we may differentiate V with respect to either x, y, or z by differentiating under the integral signs. Thus :

$$D_x V = \int\int\int D_x \left[\frac{\rho\, dx'dy'dz'}{r} \right]$$

$$= \int\int\int \frac{\rho(x'-x)\,dx'dy'dz'}{[(x'-x)^2 + (y'-y)^2 + (z'-z)^2]^{\frac{3}{2}}}, \quad [48]$$

where the limits of integration are unchanged by the differentiation. The dexter integral in this equation is (Section 14) the value of the component parallel to the axis of x of the attraction at P due to the given masses, so that we may write, using our old notation,

$$D_x V = X, \qquad [49]$$

and, similarly,

$$D_y V = Y, \qquad [50]$$

$$D_z V = Z. \qquad [51]$$

The resultant attraction at P is

$$R = \sqrt{X^2 + Y^2 + Z^2} = \sqrt{(D_x V)^2 + (D_y V)^2 + (D_z V)^2}, \quad [52]$$

and the direction-cosines of its line of action are:

$$\cos a = \frac{D_x V}{R}, \quad \cos \beta = \frac{D_y V}{R}, \quad \text{and} \quad \cos \gamma = \frac{D_z V}{R}. \quad [53]$$

It is evident from the definition of the potential function that the value of the latter at any point is independent of the particular system of rectangular axes chosen. If, then, we wish to find the component, in the direction of any line, of the attraction at any point P, we may choose one of our coördinate axes parallel to this line, and, after computing the general value of V, we may differentiate the latter partially with respect to the coördinate measured on the axis in question, and substitute in the result the coördinates of P.

21. Theorem. The results of the last section may be summed up in the words of the following

THEOREM.

To find the component at a point P, in any direction PK, of the attraction due to any attracting mass M, we may divide the difference between the values of the potential function due to M at P' (a point between P and K on the straight line PK) and at P by the distance PP'. The limit approached by this fraction as P' approaches P is the component required.

We might have arrived at this theorem in the following way:
If X, Y, Z are the components parallel to the coördinate axes of the attraction at any point P, the component in any direction PK whose direction-cosines are $\lambda, \mu,$ and ν, is

$$\lambda X + \mu Y + \nu Z = \lambda D_x V + \mu D_y V + \nu D_z V. \quad [54]$$

Let x, y, z be the coördinates of P, and $x + \Delta x, y + \Delta y, z + \Delta z$ those of P', a neighboring point on the line PK.

If V and V' are the values of the potential function at P and P' respectively, we have, by Taylor's Theorem,

$$V' = V + \Delta x \cdot D_x V + \Delta y \cdot D_y V + \Delta z \cdot D_z V + \epsilon,$$

where ϵ is an infinitesimal of an order higher than the first.

$$\frac{V'-V}{PP'} = \frac{\Delta x}{PP'} \cdot D_z V + \frac{\Delta y}{PP'} \cdot D_y V + \frac{\Delta z}{PP'} \cdot D_z V + \frac{\epsilon}{PP'} ; \quad [55]$$

but $\quad \Delta x = \lambda \cdot PP', \quad \Delta y = \mu \cdot PP', \quad \Delta z = \nu \cdot PP',$

therefore, $\underset{PP' \doteq 0}{\text{limit}} \left(\frac{V'-V}{PP'} \right) = \lambda D_z V + \mu D_y V + \nu D_z V, \qquad [56]$

and this (see [54]) is the component in the direction PK of the attraction at P: which was to be proved.

22. The Potential Function everywhere Finite. If P is a point within the attracting mass, the sum whose limit expresses the value of the potential function at P contains one apparently infinite term. That V is not infinite in this case is easily proved by making P the origin of a system of polar coördinates as in Section 15, when it will appear that the value of the potential function at P can be expressed in the form

$$V_P = \int\int\int \rho r \sin\theta \, dr \, d\theta \, d\phi ; \qquad [57]$$

and this is evidently finite.

Although V_P is everywhere finite, yet when we express its value by means of the equation [47], the quantity under the integral signs becomes infinite within the limits of integration,

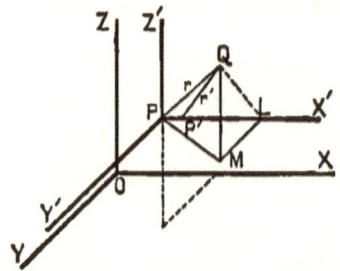

Fig. 15.

when P is a point inside the attracting mass. Under these circumstances we cannot assume without further proof that the result obtained by differentiating with respect to x under the integral signs is really $D_x V$. It is therefore desirable to compute the limit of the ratio of the difference $(\Delta_x V)$ between the values of V at the points $P' = (x + \Delta x, y, z)$ and $P = (x, y, z)$, both within the attracting mass, to the distance (Δx) between these points. For convenience, draw through P (Fig. 15) three lines parallel to the coördinate axes, and let $Q = (x', y', z')$.

Let $PQ = r$, $P'Q = r'$, and $X'PQ = \psi$.

Then
$$r'^2 = r^2 + (\Delta x)^2 - 2\,r \cdot \Delta x \cdot \cos\psi,$$

where
$$\cos\psi = \frac{x' - x}{r},$$

and
$$\frac{\Delta_z V}{\Delta x} = \iiint \left(\frac{1}{r'} - \frac{1}{r}\right) \frac{\rho\,dx'\,dy'\,dz'}{\Delta x}$$
$$= \iiint \left(\frac{r^2 - r'^2}{r'\,r^2 + rr'^2}\right) \frac{\rho\,dx'\,dy'\,dz'}{\Delta x}$$
$$= \iiint \left(\frac{2\,r\,\Delta x\,\cos\psi - (\Delta x)^2}{r'\,r^2 + rr'^2}\right) \frac{\rho\,dx'\,dy'\,dz'}{\Delta x}.$$

Therefore
$$D_z V = \lim_{\Delta x \doteq 0} \left(\frac{\Delta_z V}{\Delta x}\right)$$
$$= \iiint \frac{2\,r\,\cos\psi}{2\,r^3} \cdot \rho\,dx'\,dy'\,dz'$$
$$= \iiint \frac{\rho\,dx'\,dy'\,dz'\,\cos\psi}{r^2}. \qquad [58]$$

This last integral is evidently the component parallel to the axis of x of the attraction at P, so that the theorem of Article 21 may be extended to points within the attracting mass.

It is to be noticed that ρ is a function of x', y', and z', but not a function of x, y, and z, and that we have really proved that the derivatives with regard to x, y, and z of
$$\iiint \frac{F(x', y', z')}{r}\,dx'\,dy'\,dz',$$

where F is any finite, continuous, and single-valued function of x', y', and z', can always be found by differentiating under the integral signs, whether (x, y, z) is contained within the limits of integration or not.

23. The Potential Function due to a Straight Wire. Let μ be the mass of the unit length of a uniform straight wire AB (Fig. 16) of length $2l$. Take the middle point of the wire for the origin of coördinates, and a line drawn perpendicular to the wire at this point for the axis of x.

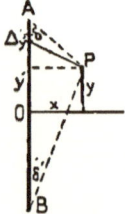

FIG. 16.

The value of the potential function at any point $P(x, y)$ in the coördinate plane is, then, according to [47],

$$V_P = \int_{-l}^{+l} \frac{\mu\, dy'}{[x^2 + (y' - y)^2]^{\frac{1}{2}}} = \mu \left[\log\{ \sqrt{x^2 + (y' - y)^2} + y' - y \} \right]_{-l}^{+l}$$

$$= \mu \log \left\{ \frac{l - y + \sqrt{x^2 + (l - y)^2}}{\sqrt{x^2 + (l + y)^2} - l - y} \right\}. \quad [59]$$

If $r = AP = \sqrt{x^2 + (l - y)^2}$, and $r' = BP = \sqrt{x^2 + (l + y)^2}$, whence $y = \dfrac{r'^2 - r^2}{4l}$, we may eliminate x and y from [59] and express V_P in terms of r and r'.

Thus:

$$V_P = \mu \log \frac{(r + 2l)^2 - r'^2}{r^2 - (r' - 2l)^2} = \mu \log \frac{r + r' + 2l}{r + r' - 2l}. \quad [60]$$

It is evident from [60] that if P move so as to keep the sum of its distances from the ends of the wire constant, V_P will remain constant. P's locus in this case is an ellipse whose foci are at A and B.

From [59] we get

$$D_z V_P = \frac{\mu}{x}\left[\frac{x^2}{r\left[r+(l-y)\right]} - \frac{x^2}{r'\left[r'-(l+y)\right]}\right]$$

$$= \frac{\mu}{x}\left[\frac{r^2-(l-y)^2}{r\left[r+(l-y)\right]} - \frac{r'^2-(l+y)^2}{r'\left[r'-(l+y)\right]}\right]$$

$$= \frac{\mu}{x}\left[\frac{r-(l-y)}{r} - \frac{r'+(l+y)}{r'}\right]$$

$$= \frac{\mu}{x}\left[1 - \cos\delta - 1 - \cos\delta'\right]$$

$$= -\frac{\mu}{x}\left[\cos\delta + \cos\delta'\right],$$

and this (Section 6). is the component in the direction of the axis of x of the attraction at P.

24. The Potential Function due to a Spherical Shell. In order to find the value at the point P of the potential function due to a homogeneous spherical shell of density ρ and of radii r_0 and r_1, we may make use of the notation of Section 9.

$$V = \iiint \frac{\rho r^2 \sin\theta\, dr\, d\theta\, d\phi}{y} = \iiint \frac{\rho r\, dy\, dr\, d\phi}{c}$$

$$= \frac{2\pi\rho}{c}\int_{r_0}^{r_1} r\, dr \int_{y_0}^{y_1} dy. \qquad [61]$$

If P lies within the cavity enclosed by the shell, the limits of y are $(r-c)$ and $(r+c)$, whence

$$V = 2\pi\rho(r_1^2 - r_0^2). \qquad [62]$$

If P lies without the shell, the limits of y are $(c-r)$ and $(c+r)$, whence

$$V = \frac{4}{3}\pi\rho\frac{(r_1^3 - r_0^3)}{c}. \qquad [63]$$

If P is a point within the mass of the shell itself. at a distance c from the centre, we may divide the shell into two parts by means of a spherical surface drawn concentric with the given shell so as to pass through P. The value of the potential function at P is the sum of the components due to these portions of the shell; therefore

$$V = 2\pi\rho(r_1^2 - c^2) + \frac{4}{3}\frac{\pi\rho}{c}(c^3 - r_0^3)$$

$$= 2\pi\rho\left\{r_1^2 - \frac{c^2}{3}\right\} - \frac{4\pi\rho}{3c}r_0^3. \qquad [64]$$

If we put these results together, we shall have the following table : —

	$c > r_0$	$r_0 < c < r_1$	$r_1 < c$
$V =$	$2\pi\rho(r_1^2 - r_0^2)$	$2\pi\rho\left(r_1^2 - \dfrac{c^2}{3}\right) - \dfrac{4\pi\rho}{3c}r_0^3$	$\dfrac{4\pi\rho}{3c}(r_1^3 - r_0^3)$
$D_c V =$	0	$\dfrac{4\pi\rho}{3}\left(\dfrac{r_0^3}{c^2} - c\right)$	$-\dfrac{4\pi\rho}{3c^2}(r_1^3 - r_0^3)$
$D_c^2 V =$	0	$-\dfrac{4\pi\rho}{3}\left(\dfrac{2r_0^3}{c^3} + 1\right)$	$\dfrac{8\pi\rho}{3c^3}(r_1^3 - r_0^3)$

If we make V, $D_c V$, and $D_c^2 V$ the ordinates of curves whose abscissas are c, we get Fig. 17.*

Here $LNQS$ represents V, and it is to be noticed that this curve is everywhere finite, continuous, and continuous in direction. The curve $OABC$ represents $D_c V$. This curve is everywhere finite and continuous, but its direction changes abruptly when the point P enters or leaves the attracting mass. The three disconnected lines OA, DE, and FG represent $D_c^2 V$.

If the density of the shell instead of being uniform were a function of the distance from the centre [$\rho = f(r)$], we should have at the point P, at the distance c from the centre of the sphere,

$$V = \frac{2\pi}{c}\int_{r_0}^{r_1} f(r) \cdot r \cdot dr \int_{y_0}^{y_1} dy. \qquad [65]$$

---

* See Thomson and Tait's *Treatise on Natural Philosophy.*

From this it follows, as the reader can easily prove, that the value of the potential function due to a spherical shell whose density is a function of the distance from the centre only is

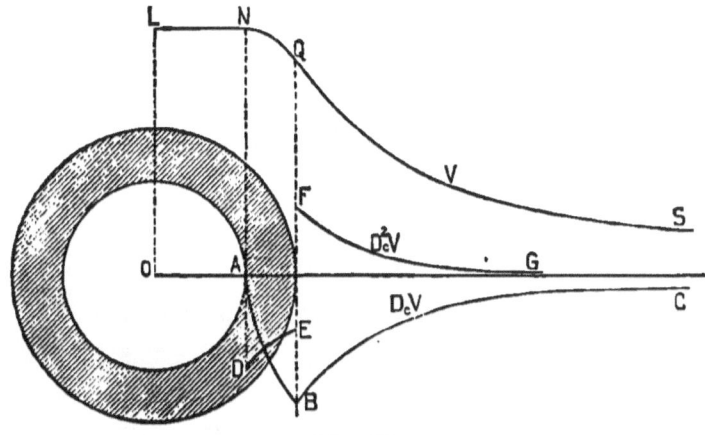

FIG. 17.

constant throughout the cavity enclosed by the shell, and at all outside points is the same as if the mass of the shell were concentrated at its centre.

25. Equipotential Surfaces. As we have already seen, V is, in general, a function of the three space coördinates $[V = f(x, y, z)]$, and in any given case all these points at which the potential function has the particular value c lie on the surface whose equation is

$$V = f(x, y, z) = c.$$

Such a surface is called an " equipotential " or " level " surface. By giving to c in succession different constant values, the equation $V = c$ yields a whole family of surfaces, and it is always possible to draw through any given point in a field of force a surface at all points of which the potential function has the same value. The potential function cannot have two different values at the same point in space, therefore no two different surfaces of the family $V = c$, where V is the potential function due to an actual distribution of matter, can ever intersect.

THEOREM.

If there be any resultant force at a point in space, due to any attracting masses, this force acts along the normal to that equipotential surface on which the point lies.

For, let $V = f(x, y, z) = c$ be the equation of the equipotential surface drawn through the point in question, and let the coördinates of this point be x_0, y_0, z_0. The equation of the plane tangent to the surface at the point is

$$(x - x_0) D_{x_0} V + (y - y_0) D_{y_0} V + (z - z_0) D_{z_0} V = 0,$$

and the direction-cosines of any line perpendicular to this plane, and hence of the normal to the given surface at the point (x_0, y_0, z_0), are

$$\cos \alpha = \frac{D_{x_0} V}{\sqrt{(D_{x_0} V)^2 + (D_{y_0} V)^2 + (D_{z_0} V)^2}}, \qquad [66_A]$$

$$\cos \beta = \frac{D_{y_0} V}{\sqrt{(D_{x_0} V)^2 + (D_{y_0} V)^2 + (D_{z_0} V)^2}}, \qquad [66_B]$$

and $\qquad \cos \gamma = \dfrac{D_{z_0} V}{\sqrt{(D_{x_0} V)^2 + (D_0 V)^2 + (D_{z_0} V)^2}} \cdot \qquad [66_C]$

But if we denote the resultant force of attraction at the point (x_0, y_0, z_0) by R, and its components parallel to the coördinate axes by X, Y, and Z, these cosines are evidently equal to $\dfrac{X}{R}$, $\dfrac{Y}{R}$, and $\dfrac{Z}{R}$ respectively, so that α, β, and γ are the direction-angles not only of the normal to the equipotential surface at the point (x_0, y_0, z_0), but also [35] of the line of action of the resultant force at the point. Hence our theorem.

Fig. 18 represents a meridian section of four of the system of equipotential surfaces due to two equal spheres whose sections are here shaded. The value of the potential function due to two spheres, each of mass M, at a point distant respectively r_1 and r_2 from the centres of the spheres, is

$$V = M\left(\frac{1}{r_1} + \frac{1}{r_2}\right),$$

and if we give to V in this equation different constant values, we shall have the equations of different members of the system of equipotential surfaces. Any one of these surfaces may be easily plotted from its equation by finding corresponding values

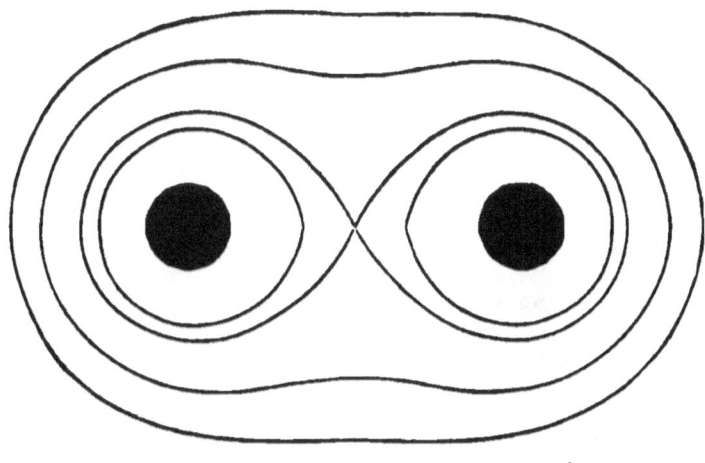

FIG. 18.

of r_1 and r_2 which will satisfy the equation; and then, with the centres of the two spheres as centres and these values as radii. describing two spherical surfaces. The intersection of these surfaces, if they intersect at all, will be a line on the surface required.

If $2a$ is the distance between the centres of the spheres.
$V = \dfrac{2M}{a}$ gives an equipotential surface shaped like an hourglass. Larger values of V than this give equipotential surfaces, each one of which consists of two separate closed ovals. one surrounding one of the spheres, and the other the other. Values of V less than $\dfrac{2M}{a}$ give single surfaces which look more and more like ellipsoids the smaller V is.

Several diagrams showing the forms of the equipotential surfaces due to different distributions of matter are given at the end of the first volume of Maxwell's *Treatise on Electricity and Magnetism.*

26. The Value of V at Infinity. The value, at the point P, of the potential function due to any attracting mass M has been defined to be

$$V = \underset{\Delta m \doteq 0}{\text{limit}} \sum \frac{\Delta m}{r}.$$

Let r_0 be the distance of the nearest point of the attracting mass from P, then

$$V < \frac{1}{r_0} \sum \Delta m \text{ or } \frac{M}{r_0}. \qquad [67]$$

The fraction $\dfrac{M}{r_0}$ has a constant numerator, and a denominator which grows larger without limit the farther P is removed from the attracting masses; hence, we see that, other things being equal, the value at P of the potential function is smaller the farther P is from the attracting matter; and that if P be moved away indefinitely, the value of the potential function at P approaches zero as a limit. In other words, *the value of the potential function at "infinity" is zero.*

27. The Potential Function as a Measure of Work. The amount of work required to move a unit mass, concentrated at a point, from one position, P_1, to another, P_2, by *any* path, in face of the attraction of a system of masses, M, is equal to

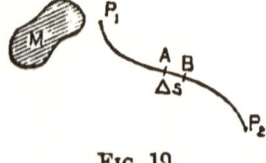

FIG. 19.

$V_1 - V_2$, where V_1 and V_2 are the values at P_1 and P_2 of the potential function due to M.

To prove this, let us divide the given path into equal parts of length Δs, and call the average force which *opposes* the motion of the unit mass on its journey along one of these elements AB (Fig. 19), F. The amount of work required to move the unit mass from A to B is $F\Delta s$, and the whole work done by moving this mass from P_1 to P_2 will be

$$\operatorname*{limit}_{\Delta s \,\doteq\, 0} \sum_{P_1}^{P_2} F\Delta s.$$

As Δs is made smaller and smaller, the average force opposing the motion along AB approaches more and more nearly the actual opposing force at A. which is $-D_s V$: therefore

$$\operatorname*{limit}_{\Delta s \,\doteq\, 0} \sum_{P_1}^{P_2} F\Delta s = -\int_{P_1}^{P_2} D_s V \cdot ds = V_1 - V_2.$$

It is to be carefully noticed that the *decrease* in the potential function in moving from P_1 to P_2 measures the work required to move the unit mass from P_1 to P_2. If P_2 is removed farther and farther from M, V_2 approaches zero, and $V_1 - V_2$ approaches V_1 as its limit. so that the value at any point P_1, of the potential function due to any system of attracting masses, is equal to the work which would be required to move a unit mass, supposed concentrated at P_1, from P_1 to "infinity" by any path.

The work (W) that must be done in order to move an attracting mass M' against the attraction of any other mass M, from a given position by any path to "infinity," is the sum of the quantities of work required to move the several elements ($\Delta m'$) into which we may divide M', and this may be written in the form

$$W = \operatorname*{limit}_{\Delta m' \,\doteq\, 0} \sum \Delta m' \int\!\!\int\!\!\int \frac{\rho\, dx\, dy\, dz}{[(x'-x)^2 + (y'-y)^2 + (z'-z)^2]^{\frac{1}{2}}}$$

$$= \int\!\!\int\!\!\int\!\!\int\!\!\int\!\!\int \frac{\rho\rho'\, dx\, dy\, dz\, dx'\, dy'\, dz'}{[(x'-x)^2 + (y'-y)^2 + (z'-z)^2]^{\frac{1}{2}}} \qquad [68]$$

W is called by some writers "the potential of the mass M' with reference to the mass M"; by others, the negative of W is called "the mutual potential energy of M and M'."

In many of the later books on this subject, the word

"potential" is never used for the value of the potential function at a point, but is reserved to denote the work required to move a mass from some present position to infinity. If V is the value of the potential function at a point P, at which a mass m is supposed to be concentrated, mV is *the potential of the mass m*. If we could have a unit mass concentrated at a point, *the potential of this mass* and the *value of the potential function at the point* would be numerically identical.

28. Laplace's Equation. We have seen that the value of the potential function and the component in any direction of the attraction at the point P are always finite functions of the space coördinates, whether P is inside, outside, or at the surface of the attracting masses. We have seen also that by differentiating V at any point with respect to any direction we may find the always finite component in that direction of the attraction at the point. It follows that D_xV, D_yV, D_zV are everywhere finite, and that, in consequence of this, the potential function is everywhere continuous as well as finite.

If P is a point outside of the attracting masses, the quantity under the integral signs in [48], by which $dx'dy'dz'$ is multiplied, cannot be infinite within the limits of integration, and we can find D_x^2V by differentiating the expression for D_xV under the integral signs.

In this case

$$D_x^2V = \int\int\int \frac{3(x'-x)^2 - r^2}{r^5} \rho\, dx'dy'dz', \qquad [69]$$

and similarly,

$$D_y^2V = \int\int\int \frac{3(y'-y)^2 - r^2}{r^5} \rho\, dx'dy'dz', \qquad [70]$$

$$D_z^2V = \int\int\int \frac{3(z'-z)^2 - r^2}{r^5} \rho\, dx'dy'dz'. \qquad [71]$$

Whence, for all points exterior to the attracting masses,

$$D_x^2V + D_y^2V + D_z^2V = 0. \qquad [72]$$

This is Laplace's Equation.

The operator $(D_x^2 + D_y^2 + D_z^2)$ is sometimes denoted by the symbol ∇^2, so that [72] may be written

$$\nabla^2 V = 0. \qquad [73]$$

The potential function, due to every conceivable distribution of matter, must be such that at all points in empty space Laplace's Equation shall be satisfied.

29. The Second Derivatives of the Potential Function are Finite at Points within the Attracting Mass. If the point P lies within the attracting mass, V and $D_x V$ are finite, but the quantity under the integral signs in the expression for $D_x V$ becomes infinite within the limits of integration, and we cannot assume that $D_x^2 V$ may be found by differentiating $D_x V$ under the integral signs. In order to find $D_x^2 V$ under these circumstances, it is convenient to transform the equation for $D_x V$. Let us choose our coördinate axes so as to have all the attracting mass in the first octant, and divide the projection of the contour of this mass on the plane yz into elements $(dy'dz')$. Upon each one of these elements let us erect a right prism, cutting the mass twice or some other even number of times. Consider one of the elements $dy'dz'$ whose corner next the origin has the coördinates 0, y', and z'. The prism erected on this element cuts out elements ds_1, ds_2, ds_3, ds_4, $\cdots ds_{2n}$ from the surface of the attracting mass and that edge of the prism which is perpendicular to the plane yz at $(0, y', z')$ cuts into the surface at points whose distances from the plane of yz are $a_1, a_3, a_5, \cdots a_{2n-1}$, and out of the surface at points whose distances from the same plane are $a_2, a_4, a_6, \cdots a_{2n}$. At every one of these points of intersection draw normals towards the interior of the attracting mass, and call the angles which these normals make with the positive direction of the axis of x, $a_1, a_2, a_3, \cdots a_{2n}$. It is to be noticed that $a_1, a_3, a_5, \cdots a_{2n-1}$ are all acute, and that $a_2, a_4, a_6, \cdots a_{2n}$ are all obtuse. The element $dy'dz'$ may be regarded as the common projection of the surface elements

$ds_1,\ ds_2,\ ds_3,\ \cdots\ ds_{2n}$, and, so far as absolute value is concerned, the following equations hold approximately:

$$dy'dz' = ds_1 \cos a_1 = ds_2 \cos a_2 = ds_3 \cos a_3 = \cdots = ds_{2n} \cos a_{2n}.$$

But $dy'dz'$, ds_1, ds_2, ds_3, etc., are all positive areas, and $\cos a_2$, $\cos a_4$, $\cos a_6$, etc., are negative, so that, paying attention to signs as well as to absolute values, we have

$$dy'dz' = +ds_1 \cos a_1 = -ds_2 \cos a_2 = +ds_3 \cos a_3 = -ds_4 \cos a_4 = \text{etc.}$$

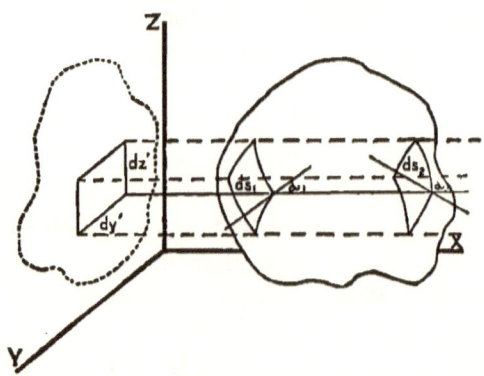

FIG. 20.

Now

$$D_x V = \iiint \frac{\rho(x'-x)\,dx'dy'dz'}{r^3} = \iint dy'dz' \int \rho\, D_{x'}\left(-\frac{1}{r}\right) dx',\quad [74]$$

and in order to find the value of this expression by the use of the prisms just described, we are to cut each one of these prisms into elementary rectangular parallelopipeds by planes parallel to the plane of yz; we are to multiply the values of every one of these elements which lies within the attracting mass by the value of $\rho\, D_{x'}\left(-\dfrac{1}{r}\right)$ at its corner next the origin [*i.e.*, at (x',y',z')]; and we are to find the limit of the sum of these as dx' is made smaller and smaller. We are then to compute a like expression for each of the other prisms, and to find the limit of the sum of the whole as the bases of the prisms are made smaller and smaller and their number correspondingly increased.

Wherever the function $\frac{\rho}{r}$ is a continuous and finite function of x', we have

$$D_z'\frac{\rho}{r} = \frac{1}{r}D_z'\rho + \rho D_z'\frac{1}{r} = \frac{1}{r}D_z'\rho - \rho D_z'\left(-\frac{1}{r}\right):$$

hence, if the elementary prisms cut the surface of the attracting mass only twice,

$$D_z V = \int\int dy' dz' \left[-\frac{\rho}{r}\right]_{x'=a_1}^{x'=a_2} + \int\int\int \frac{1}{r}D_z'\rho\, dx' dy' dz' \; ; \quad [75]$$

and, in general,

$$D_z V = \int\int dy'\, dz' \left[\frac{\rho_1}{r_1} - \frac{\rho_2}{r_2} + \frac{\rho_3}{r_3} - \frac{\rho_4}{r_4} + \cdots - \frac{\rho_{2n}}{r_{2n}}\right]$$

$$+ \int\int\int \frac{1}{r}D_z'\rho\, dx'\, dy'\, dz' \qquad [76]$$

$$= \lim \sum \left(\frac{\rho_1}{r_1}\cos a_1\, ds_1 + \frac{\rho_2}{r_2}\cos a_2\, ds_2 + \frac{\rho_3}{r_3}\cos a_3\, ds_3 + \cdots\right.$$

$$\left. + \frac{\rho_{2n}}{r_{2n}}\cos a_{2n}\, ds_{2n}\right) + \int\int\int \frac{1}{r}D_z'\rho\, dx'\, dy'\, dz', \qquad [77]$$

where $\frac{\rho_k}{r_k}$ is the value of the quantity $\frac{\rho}{r}$ at the point where the line $y = y'$, $z = z'$ cuts the surface of the attracting mass for the kth time, counting from the plane yz.

In order to find the value of the limit of the sum which occurs in this expression, it is evident that we may divide the *entire surface* of the attracting mass into elements, multiply the area of each element by the value of $\frac{\rho\cos a}{r}$ at one of its points, and find the limit of the sum formed by adding all these products together; but this is equivalent to the surface integral of $\frac{\rho\cos a}{r}$ taken all over the outside of the attracting mass, so that

$$D_z V = \int \frac{\rho}{r}\cos a\, ds + \int\int\int \frac{D_z'\rho}{r}\, dx'\, dy'\, dz', \qquad [78]$$

where the first integral is to be taken all over the surface of the attracting mass and the second throughout its volume. This expression for $D_z V$ is in some cases more convenient than that of [48].

We have proved this transformation to be correct, however, only when $\frac{\rho}{r}$ is finite throughout the attracting mass. If P is a point within the mass, $\frac{\rho}{r}$ is infinite at P. In this case surround P by a spherical surface of radius ϵ small enough to make the whole sphere enclosed by this surface lie entirely within the attracting mass. This is possible unless P lies exactly upon the surface of the attracting mass. Shutting out the little sphere, let V_2 be the potential function due to the rest (T_2) of the attracting mass; then, since P is an outside point with regard to T_2, we have, by [78],

$$D_x V_2 = \int \frac{\rho}{\epsilon} \cos a \cdot ds' + \int \frac{\rho}{r} \cos a\, ds + \int\int\int \frac{D_z' \rho}{r} dx'dy'dz', \quad [79]$$

where the first integral is to be extended over the spherical surface, which forms a part of the boundary of the attracting

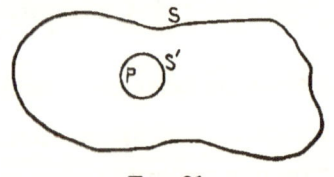

Fig. 21.

mass to which V_2 is due; the second integral is to be taken over all the rest of the bounding surface of the attracting mass; and the triple integral embraces the volume of all the attracting mass which gives rise to V_2.

As ϵ is made smaller and smaller, V_2 approaches more and more nearly the potential function V, due to all the attracting mass.

In the integral $\int \frac{\rho}{\epsilon} \cos a\, ds'$, $\cos a$ can never be greater than 1 nor less than -1, so that if $\bar{\rho}$ is the greatest value of ρ on the surface of the sphere, the absolute value of the integral must be less than $\dfrac{\bar{\rho}}{\epsilon} \int^\bullet ds'$ or $4\pi\bar{\rho}\epsilon$, and the limit of this as ϵ approaches zero is zero. The second integral in [79] is unaltered by any change in ϵ. If we make P the origin of a system of polar coördinates, it is evident that the triple integral in [79] may be written

$$\iiint D_x'\rho \cdot r\sin\theta\, dr\, d\theta\, d\phi, \qquad [80]$$

and the limit which this approaches as ϵ is made smaller and smaller is evidently finite, for, if $r = 0$, the quantity under the integral sign is zero.

Therefore,

$$\operatorname*{limit}_{\epsilon \doteq 0} D_x V_2 = D_x V = \int \frac{\rho}{r}\cos a\, ds + \iiint \frac{D_x'\rho}{r} dx'\, dy'\, dz', \quad [81]$$

and [79] is true even when P lies within the attracting mass.

Under the same conditions we have, similarly,

$$D_y V = \int^\bullet \frac{\rho}{r}\cos\beta\, ds + \iiint \frac{D_y'\rho}{r}\, dx'\, dy'\, dz', \qquad [82]$$

and

$$D_z V = \int^\bullet \frac{\rho}{r}\cos\gamma\, ds + \iiint \frac{D_z'\rho}{r}\, dx'\, dy'\, dz'. \qquad [83]$$

Observing that in these surface integrals r can never be zero, since we have excluded the case where P lies on the surface of the attracting mass, and that the triple integrals belong to the class mentioned in the latter part of Section 22, we will differentiate [81], [82], and [83] with respect to x, y, and z respectively, by differentiating under the integral signs. If the results are finite, we may consider the process allowable.

Performing the work indicated, we have

$$D_x^2 V = \int \rho \cos a \cdot D_x\left(\frac{1}{r}\right) ds + \iiint D_x\left(\frac{1}{r}\right) \cdot D_x'\rho \cdot dx'\, dy'\, dz'. [84]$$

$$D_y^2 V = \int \rho \cos \beta \cdot D_y\left(\frac{1}{r}\right) ds + \iiint D_y\left(\frac{1}{r}\right) \cdot D_y'\rho \cdot dx'\, dy'\, dz'. [85]$$

$$D_z^2 V = \int \rho \cos \gamma \cdot D_z\left(\frac{1}{r}\right) ds + \iiint D_z\left(\frac{1}{r}\right) \cdot D_z'\rho \cdot dx'\, dy'\, dz'. [86]$$

and by making P the centre of a system of polar coördinates and transforming all the triple integrals, it is easy to show that the values of $D_x^2 V$, $D_y^2 V$, $D_z^2 V$ here found are finite whether P is within or without the attracting mass. This result * is important.

30. The Derivatives of the Potential Function at the Surface of the Attracting Mass.

Let the point P lie on the surface of the attracting mass, or at some other point or surface where ρ is discontinuous. Make P the centre of a sphere of radius ϵ, and call the piece which this sphere cuts out of the attracting mass T_1 and the remainder of this mass T_2. Let V_1 and V_2 be the potential functions due respectively to T_1 and T_2, then

$$V = V_1 + V_2, \quad D_x V = D_x V_1 + D_x V_2,$$

and the increment $[\Delta(D_x V)]$ made in $D_x V$ by moving from P to a neighboring point P', inside T_1, is equal to the sum of the corresponding increments $[\Delta(D_x V_1)$ and $\Delta(D_x V_2)]$ made in $D_x V_1$ and $D_x V_2$.

With reference to the space T_2, P is an outside point, so that the values at P of the first derivatives of V_2 with respect to x, y, and z are continuous functions of the space coördinates and

$$\underset{PP' \doteq 0}{\text{limit}} \Delta(D_x V_2) = 0.$$

Let $d\omega$ be the solid angle of an elementary cone whose vertex is at any fixed point O in T_1 used as a centre of coördinates.

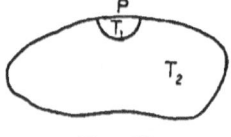

Fig. 22.

The element of mass will be $\rho r^2 d\omega dr$. The component in the direction of the axis of x of the attraction at O due to T_1 is the

---

* Lejeune Dirichlet, *Vorlesungen über die im umgekehrten Verhältniss des Quadrats der Entfernung wirkenden Kräfte.*

Riemann, *Schwere, Electricität, und Magnetismus.*

limit of the sum taken throughout T_1 of $\dfrac{\rho r^2 a\, d\omega\, dr}{r^2}$, where a is the cosine of the angle which the line joining O with the element in question makes with the axis of x. The difference between the limits of ω is not greater than 4π, and the difference between the limits of r is not greater than 2ϵ. If, then, κ is the greatest value which ρa has in T_1,

$$(D_x V_1)_0 < 8\pi\kappa\epsilon.$$

It follows from this that if P' is a point within T_1 so that $PP' < \epsilon$, the change made in $D_x V_1$ by going from P to P' is far less than $16\pi\kappa\epsilon$; but this last quantity can be made as small as we like by making ϵ small enough, so that

$$\underset{PP' \doteq 0}{\text{limit}}\, \Delta(D_x V_1) = 0,$$

whence

$$\underset{PP' \doteq 0}{\text{limit}}\, \Delta(D_x V) = \underset{PP' \doteq 0}{\text{limit}}\, \Delta(D_x V_1) + \underset{PP' \doteq 0}{\text{limit}}\, \Delta(D_x V_2) = 0,$$

and $D_x V$ varies continuously in passing through P. In a similar manner, it may be proved that $D_y V$ and $D_z V$ are everywhere, even at places where the density is discontinuous, continuous functions of the space coördinates.

The results of the work of the last two sections are well illustrated by Fig. 17. We might prove, with the help of a transformation due to Clausius,[*] that the second derivatives of the potential function are finite at all points on the surface of the attracting matter where the curvature is finite, but that these derivatives generally change their values abruptly whenever the point P crosses a surface at which ρ is discontinuous, as at the surface of the attracting masses. The fact, however, that this last is true in the special case of a homogeneous spherical shell suffices to show that we cannot expect the second derivatives of V to have definite values at the boundaries of attracting bodies.

---

[*] *Die Potentialfunction und das Potential*, §§ 19–24.

31. Gauss's Theorem. If any closed surface T drawn in a field of force be divided up into a large number of surface elements, and if each one of these elements be multiplied by the component, in the direction of the interior normals of the force of attraction at a point of the element, and if these products be added together, the limit of the sum thus obtained is called the " surface integral of normal attraction over T."

If any closed surface T be described so as to shut in completely a mass m concentrated at a point, the surface integral of normal attraction due to m, taken over T, is $4\pi m$; and, in general, if any closed surface T be described so as to shut in completely any system of attracting masses M, the surface integral over T of the normal attraction due to M is $4\pi M$.

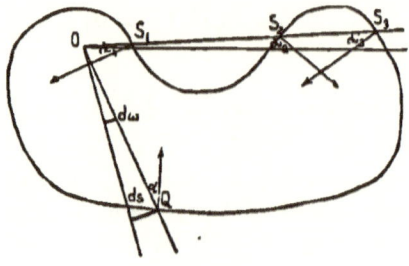

FIG. 23.

In order to prove this, divide T up into surface elements, and consider one of these ds at Q. The attraction at Q in the direction QO, due to the mass m concentrated at O, is $\dfrac{m}{QO^2} = \dfrac{m}{r^2}$. The component of this in the direction of the interior normal is $\dfrac{m}{r^2}\cos a$, and the contribution which ds yields to the sum whose limit is the surface integral required is $\dfrac{m\cos a\,ds}{r^2}$. Connect every point of the perimeter of ds with O by a straight line, thus forming a cone of such size as to cut out of a spherical surface of unit radius drawn about O an element $d\omega$, say. If we draw about O a sphere of radius $r = OQ$, the cone will intercept on its surface an element equal to $r^2 \cdot d\omega$. This element is the projection on the spherical surface of ds; hence $ds \cos a = r^2 d\omega$, approximately, and the contribution of the element ds to our surface integral is $m\,d\omega$. But an elementary cone may cut the surface more than once; indeed, any odd number of times. Consider such a cone, one element of which cuts the surface thrice in S_1, S_2, and S_3. Let OS_1, OS_2, and OS_3 be called r_1, r_2, and r_3 respectively, and let the surface elements cut out of T by the cone be ds_1, ds_2, and ds_3, and the angles between the line S_3O and the interior normals to T at S_1, S_2, and S_3 be a_1, a_2, a_3. It is to be noticed that when the cone cuts out of T, the corresponding angle is acute, and that when it cuts in, the corresponding angle is obtuse. a_1 and a_3 are acute, and a_2 obtuse. If we draw about O three spherical surfaces with radii r_1, r_2, and r_3 respectively, the cone will cut out of these the elements $r_1^2 d\omega$, $r_2^2 d\omega$, and $r_3^2 d\omega$. In absolute size, $ds_1 = r_1^2 d\omega \sec a_1$, $ds_2 = r_2^2 d\omega \sec a_2$, and $ds_3 = r_3^2 d\omega \sec a_3$, approximately, but ds_2 and $r^2 d\omega$ are both positive, being areas, and $\sec a_2$ is negative. Taking account of sign, then, $ds_2 = -r^2 d\omega \sec a_2$, and the cone's three elements yield to the surface integral of normal attraction the quantity

$$\left(m \frac{ds_1 \cos a_1}{r_1^2} + \frac{ds_2 \cos a_2}{r_2^2} + \frac{ds_3 \cos a_3}{r_3^2} \right) = m\,(d\omega - d\omega + d\omega) = m\,d\omega.$$

However many times the cone cuts T, it will yield $m\,d\omega$ to the surface integral required: all such elementary cones will yield then $m \sum d\omega = m\,4\pi$ if T is closed, and, in general, $m\Theta$, when Θ is the solid angle which T subtends at O.

If, instead of a mass concentrated at a point, we have any distribution of masses, we may divide these into elements, and apply to each element the theorem just proved; hence our general statement.

If from a point O without a closed surface T an elementary cone be drawn, the cone, if it cuts T at all, will cut it an even number of times. Using the notation just explained, the contribution which any such cone will yield to the surface integral taken over T of a mass m concentrated at O is

$$m\left(\frac{ds_1\cos a_1}{r_1^2} + \frac{ds_2\cos a_2}{r_2^2} + \frac{ds_3\cos a_3}{r_3^2} + \frac{ds_4\cos a_4}{r_4^2} + \cdots\right)$$

$$= m(-d\omega + d\omega - d\omega + d\omega - \cdots) = m \cdot 0 = 0,$$

and the surface integral over any closed surface of the normal attraction due to any system of outside masses is zero.

The results proved above may be put together and stated in the form of a

THEOREM DUE TO GAUSS.

If there be any distribution of matter partly within and partly without a closed surface T, and if M be the sum of the masses which T encloses, and M′ the sum of the masses outside T, the surface integral over T of the normal attraction N toward the interior, due to both M and M′, is equal to $4\pi M$. If V be the potential function due to both M and M′, we have

$$\int N ds = \int D_n V \cdot ds = 4\pi M.$$

It is easy to see that if a mass M be supposed concentrated on the surface of any closed surface T whose curvature is everywhere finite, the surface integral of normal attraction taken over T will be $2\pi M$; for all the elementary cones which can be drawn from a point P in the surface so as to cut T once or some other odd number of times lie on one side of the tangent plane at the point, and intercept just half the surface of the sphere of unit radius whose centre is P.

From Gauss's Theorem it follows immediately that at some parts of a closed surface situated in a field of force, but enclosing none of the attracting mass, the normal component of the resultant attraction must act towards the interior of the surface and at some parts toward the exterior, for otherwise the limit of the sum of the intrinsically positive elements of the surface, each one multiplied by the component in the direction of the interior normal of the attraction at one of its own points, could not be zero. In other words, the potential function, whose rate of change measures the attraction, must at some parts of the surface increase and at others decrease in the direction of the interior normal.

From this it follows that the potential function cannot have a maximum or a minimum value at a point in empty space; for if at such a point Q the potential function had a maximum value, we could surround Q by a small closed surface, at every point of which the potential function would increase in the direction of the interior normal, and this would be inconsistent with the fact that the surface integral of normal attraction taken over the surface, which would contain no matter, must be zero. Similarly it may be shown that the potential function cannot have a minimum value at a point in empty space.

If the potential function be constant over a closed surface which contains none of the attracting mass, it has the same value throughout the interior; for if this were not the case, some point or region Q within T would have a value greater or less than the surrounding region, and we could enclose Q by a closed surface to which we could apply the course of reasoning just used to show that V cannot attain a maximum value at a point in empty space.

32. Tubes of Force. A line which cuts orthogonally the different members of the system of equipotential surfaces corresponding to any distribution of matter is called a "line of force," since its direction at each point of its course shows the direction of the resultant force at the point. If through all points of the contour of a portion of an equipotential surface lines of force be drawn, these lines lie on a surface called a

Fig. 24.

"tube of force." We may easily apply Gauss's Theorem to a space cut out and bounded by a portion of a tube of force and two equipotential surfaces; for the sides of the tube do not contribute anything to the surface integral of normal attraction, and the resultant force is all normal at points in the equipotential surfaces. If ω and ω' are the areas of the sections of a tube of force made by two equipotential surfaces, and if F and F' are the average interior forces on ω and ω', we have

$$F\omega + F'\omega' = 0 \qquad [87]$$

if the tube encloses empty space, and

$$F\omega + F'\omega' = 4\pi m \qquad [88]$$

when the tube encloses a mass m of attracting matter.

33. Spherical Distributions. In the case of a distribution about a point in spherical shells, so that the density is a function of the distance from this point only, the lines of force are straight lines whose directions all pass through the central point. Every tube of force is conical, and the areas cut out of different equipotential surfaces by a given tube of force are proportional to the square of the distance from the centre.

Consider a tube of force which intercepts an area ψ from a spherical surface of unit radius drawn with O as a centre, and apply Gauss's Theorem to a box cut out of this tube by two equipotential surfaces of radii r and $(r + \Delta r)$ respectively.

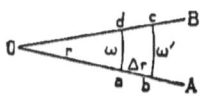

Fig. 25.

Let AOB (Fig. 25) be a section of the tube in question. The area of the portion of spherical surface ω which is represented in section at ad is $r^2\psi$, and the area of that at bc is $(r + \Delta r)^2\psi$. If the average force acting on ω toward the inside of the box is F, the average force acting on ω' toward the inside of the box will be $-(F + \Delta_r F)$, and the surface integral of normal attraction taken all over the outside of the box is

$$F r^2\psi - (F + \Delta_r F)(r + \Delta r)^2\psi = -\psi \cdot \Delta_r(F \cdot r^2). \qquad [89]$$

If the tube of force which we have been considering be extended far enough, it will cut all the concentric layers of matter, traverse all the empty regions between the layers, if there are such, and finally emerge into outside space.

If we choose r so that the box shall contain no matter, the surface integral taken over the box must be zero.

In this case,

$$-\psi \Delta_r (F r^2) = 0,$$

therefore,

$$F = \frac{c}{r^2}, \tag{90}$$

and

$$V = -\frac{c}{r} + \mu. \tag{91}$$

From this it follows that in a region of empty space, either included between the two members of a system of concentric spherical shells of density depending only upon the distance from the centre, or outside the whole system, the force of attraction at different points varies inversely as the squares of the distances of these points from the centre.

Suppose that the box ($abcd$) lies in a shell whose density is constant; then the surface integral of normal attraction taken over the box is equal to 4π times the matter within the box. In this case the quantity of matter inside the box is

$$\rho \tfrac{4}{3}\pi[(r+\Delta r)^3 - r^3]\frac{\psi}{4\pi} \quad \text{or} \quad \rho \psi r^2 \Delta r + \epsilon,$$

where ϵ is an infinitesimal of an order higher than the first. Therefore,

$$-\psi \Delta_r (F r^2) = 4\pi (\rho \psi r^2 \Delta r + \epsilon),$$

or

$$\underset{\Delta r \doteq 0}{\text{limit}} \frac{\Delta_r (F r^2)}{\Delta r} = -4\pi \rho r^2,$$

whence

$$F = -\frac{4\pi \rho r}{3} + \frac{c}{r^2}, \tag{92}$$

and

$$V = -\frac{c}{r} - \frac{2}{3}\pi \rho r^2 + \mu. \tag{93}$$

If the box lies in a shell whose density is inversely proportional to the distance from the centre, we shall have

$$\underset{\Delta r \doteq 0}{\text{limit}} \frac{\Delta_r (Fh^2)}{\Delta r} = -4\pi\left(\frac{\lambda}{r}\right)r^2, \qquad [94]$$

whence

$$F = -2\pi\lambda + \frac{c}{r^2}, \qquad [95]$$

and

$$V = -\frac{c}{r} - 2\pi\lambda r + \mu. \qquad [96]$$

In general, if the box lies in a shell whose density is $f(r)$, we shall have

$$\underset{\Delta r \doteq 0}{\text{limit}} \frac{\Delta_r (Fr^2)}{\Delta r} = -4\pi f(r)r^2, \qquad [97]$$

whence

$$F = \frac{c}{r^2} - \frac{4\pi}{r^2}\int f(r) r^2 \cdot dr. \qquad [98]$$

In order to learn how to use the results just obtained to determine the force of attraction at any point due to a given spherical distribution, let us consider the simple case of a single shell, of radii 4 and 5, and of density $[\lambda r]$ proportional to the distance from the centre.

At points within the cavity enclosed by the shell we must have, according to [90] and [91],

$$F = \frac{c}{r^2} \quad \text{and} \quad V = -\frac{c}{r} + \mu \ ;$$

But the force is evidently zero at the centre of the shell, where r is zero, so that c must be zero everywhere within the cavity, and there is no resultant force at any point in the region. The value, at the centre, of the potential function due to the shell is evidently

$$\mu = \int_4^5 4\pi\lambda r^2 dr = \frac{244\pi\lambda}{3}, \qquad [99]$$

and it has the same value at all other points in the cavity.

In the shell itself it is easy to see that we must have at all points,

$$F = \frac{c'}{r^2} - \pi\lambda r^2 \quad \text{and} \quad V = -\frac{c'}{r} - \frac{\pi\lambda r^3}{3} + \mu'. \qquad [100]$$

In order to determine the constants in this equation, we may make use of the fact that F and V are everywhere continuous functions of the space coördinates, so that the values of F and V obtained by putting $r = 4$, the inner radius of the shell, in [100], must be the same as those obtained by making $r = 4$ in the expressions which give the values of F and V for the cavity enclosed by the shell. This gives us

$$c' = 256\,\pi\lambda \quad \text{and} \quad \mu' = \frac{500\,\pi\lambda}{3},$$

so that for points within the mass of the shell we have

$$F = \frac{256\,\pi\lambda}{r^2} - \pi\lambda r^2. \qquad [101]$$

and

$$V = -\frac{256\,\pi\lambda}{r} - \frac{\pi\lambda r^3}{3} + \frac{500\,\pi\lambda}{3}. \qquad [102]$$

For points without the shell we have the same general expressions for F and V as for points within the cavity enclosed by the shell, namely,

$$F = \frac{k}{r^2} \quad \text{and} \quad V = -\frac{k}{r} + m, \qquad [103]$$

but the constants are different for the two regions.

Keeping in mind the fact that F and V are continuous, it is easy to see that we must get the same result at the boundary of the shell, where $r = 5$, whether we use [103], or [101] and [102].

This gives

$$k = -369\,\pi\lambda \quad \text{and} \quad m = 0 ;$$

so that for all points outside the shell we have

$$F = -\frac{369\,\pi\lambda}{r^2}. \qquad [104]$$

and

$$V = \frac{369\,\pi\lambda}{r}. \qquad [105]$$

These last results agree with the statements made in Section 13, for the mass of the shell is $369\,\pi\lambda$.

The values, at every point in space, of the potential function and of the attraction due to any spherical distribution may be found by determining, first, with the aid of Gauss's Theorem, the general expressions for F and V in the several regions; then the constants for the innermost region, remembering that there is no resultant attraction at the centre of the system; and finally, in succession (moving from within outwards), the constants for the other regions, from a consideration of the fact that no abrupt change in the values of either F or V is made by crossing the common boundary of two regions.

This method of treating problems is of great practical importance.

34. Cylindrical Distributions. In the case of a cylindrical distribution about an axis, where the density is a function of the distance from the axis only, the equipotential surfaces are concentric cylinders of revolution; the lines of force are straight lines perpendicular to the axis; and every tube of force is a wedge.

If we apply Gauss's Theorem to a box shut in between two equipotential surfaces of radii r and $r + \Delta r$, two planes perpendicular to the axis, and two planes passing through the axis,

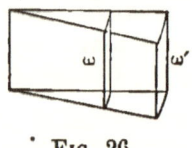

FIG. 26.

we have, if ψ is the area of the piece cut out of the cylindrical surface of unit radius by our tube of force,

$$\omega = r \cdot \psi, \quad \omega' = (r + \Delta r) \cdot \psi,$$

and for the surface integral of normal attraction taken over the box,

$$F\omega + F'\omega' = -\psi \cdot \Delta_r(r \cdot F). \qquad [106]$$

If our box is in empty space,

$$\Delta_r(r \cdot F) = 0,$$

so that $$F = \frac{c}{r} \quad \text{and} \quad V = c \log r + \mu. \qquad [107]$$

If the box is within a shell of constant density ρ,

$$-\psi \cdot \Delta_r (r \cdot F) = 4\pi\psi \rho r \Delta r,$$

so that $\quad F = \dfrac{c}{r} - 2\pi\rho r \quad$ and $\quad V = c \log r - \pi\rho r^2 + \mu.$ [108]

35. Poisson's Equation. Let us now apply Gauss's Theorem to the case where our closed surface is that of an element of volume of an attracting mass in which ρ is either constant or a continuous function of the space coördinates. We will consider three cases, using first rectangular coördinates, then cylinder coördinates, and finally spherical coördinates.

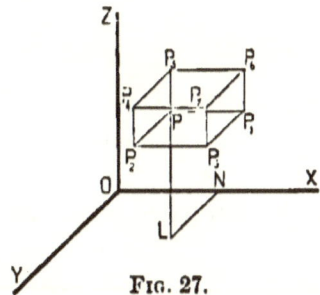

FIG. 27.

I. In the first case, our element is a rectangular parallelopiped (Fig. 27). Perpendicular to the axis of x are two equal surfaces of area $\Delta y \cdot \Delta z$, one at a distance x from the plane yz, and one at a distance $x + \Delta x$ from the same plane. The average force perpendicular to a plane area of size $\Delta y \Delta z$, parallel to the plane yz, and with edges parallel to the axes of y and z, is evidently some function of the coördinates of the corner of the element nearest the origin.

That is, if $P = (x, y, z)$, the average force on PP_4 parallel to the axis of x is $X = f(x, y, z)$, and the average force on $P_1 P_5$ in the same direction is $f(x + \Delta x, y, z) = X + \Delta_x X$, so that PP_4 and $P_1 P_5$ yield towards the surface integral of interior-normal attraction taken over the element, the quantity $-\Delta x \Delta y \Delta z \cdot \dfrac{\Delta_x X}{\Delta x}$.

Similarly, the other two pairs of elementary surfaces yield

$-\Delta x \Delta y \Delta z \dfrac{\Delta_y Y}{\Delta y}$ and $-\Delta x \Delta y \Delta z \dfrac{\Delta_z Z}{\Delta z}$, and, if ρ_0 is the average density of the matter enclosed by the box, we have

$$-\Delta x \Delta y \Delta z \left[\frac{\Delta_z X}{\Delta x_,} + \frac{\Delta_y Y}{\Delta y} + \frac{\Delta_z Z}{\Delta z} \right] = 4 \pi \rho_0 \Delta x \Delta y \Delta z. \quad [109]$$

This equation is true whatever the size of the element $\Delta x \Delta y \Delta z$. If this element is made smaller and smaller, the average normal force $[X]$ on PP_4 approaches in value the force $[D_x V]$ at P in the direction of the axis of x; Y and Z approach respectively the limits $D_y V$ and $D_z V$; and ρ_0 approaches as its limit the actual density $[\rho]$ at P.

Taking the limits of both sides of [109], after dividing by $\Delta x \Delta y \Delta z$, we have

$$D_x^2 V + D_y^2 V + D_z^2 V = - 4 \pi \rho,$$

or $\qquad \qquad \nabla^2 V = - 4 \pi \rho, \qquad\qquad\qquad [110]$

which is Poisson's Equation. The potential function due to any conceivable distribution of attracting matter must be such that at all points within the attracting mass this equation shall be satisfied.

For points in empty space $\rho = 0$, and Poisson's Equation degenerates to Laplace's Equation.

II. In the case of cylindrical coördinates, the element of volume (Fig. 28) is bounded by two cylindrical surfaces of revo-

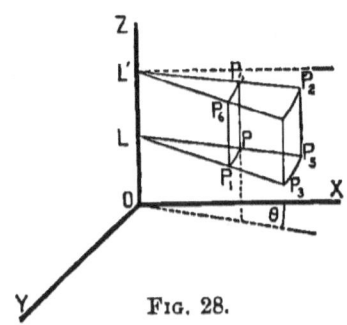

FIG. 28.

lution having the axis of z as their common axis and radii r and $r + \Delta r$, two planes perpendicular to this axis and distant Δz

from each other, and two planes passing through the axis and forming with each other the diedral angle $\Delta\theta$.

Call R, Θ, and Z the average normal forces upon the elementary planes PP_6, PP_2, and PP_3 respectively, then the surface integral of normal attraction over the volume element will be

$$- \Delta\theta\, \Delta z\, \Delta_r(r\cdot R) - \Delta r\, \Delta z\, \Delta_\theta\Theta - \Delta\theta\left[r\Delta r + \tfrac{1}{2}(\Delta r)^2\right]\Delta_z Z$$
$$= 4\pi\rho_0 \text{ (vol. of box) };\qquad\qquad [111]$$

whence, approximately,

$$\frac{1}{r}\frac{\Delta_r(rR)}{\Delta r} + \frac{1}{r}\frac{\Delta_\theta\Theta}{\Delta\theta} + \frac{\Delta_z Z}{\Delta z} = -4\pi\rho_0\frac{\text{vol. of box}}{r\,\Delta r\,\Delta\theta\,\Delta z}. \quad [112]$$

The force at P in direction PP_5 is $D_r V$, in direction PP_4 is $D_z V$, and perpendicular to LP in the plane PLP_1 is $\frac{1}{r}\cdot D_\theta V$, so that if the box is made smaller and smaller, our equation approaches the form

$$\frac{1}{r}D_r(r\cdot D_r V) + \frac{1}{r^2}D_\theta^2 V + D_z^2 V = -4\pi\rho. \quad [113]$$

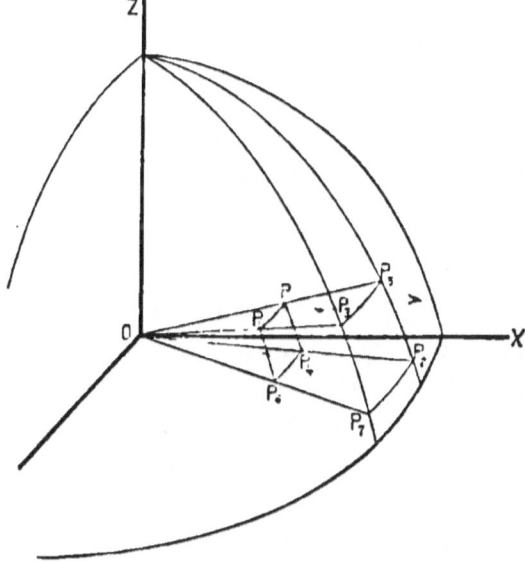

FIG. 29.

III. In the case of spherical coördinates, the volume element is of the shape shown in Fig. 29. Let $OP = r$, $ZOP = \theta$, and denote by ϕ the diedral angle between the planes ZOP and ZOX. Denote by R, Θ, and Φ the average normal forces on the faces PP_6, PP_3, and PP_2 respectively; then the surface integral of normal attraction over the elementary box is approximately

$$-\sin\theta\Delta\theta\Delta\phi\cdot\Delta_r(r^2R) - r\Delta\theta\Delta r\Delta_\phi\Phi - r\Delta\phi\Delta r\cdot\Delta_\theta(\sin\theta\cdot\Theta)$$

$$= 4\pi\rho_0\cdot(\text{vol. of box}); \qquad [114]$$

whence $\quad \dfrac{1}{r^2}\cdot\dfrac{\Delta_r(r^2R)}{\Delta r} + \dfrac{1}{r\sin\theta}\cdot\dfrac{\Delta_\phi\Phi}{\Delta\phi} + \dfrac{1}{r\sin\theta}\cdot\dfrac{\Delta_\theta(\sin\theta\cdot\Theta)}{\Delta\theta}$

$$= -4\pi\rho_0\cdot\dfrac{\text{vol. of box}}{r^2\sin\theta\Delta r\,\Delta\theta\Delta\phi}. \qquad [115]$$

The force at P in the direction PP_5 is D_rV, in the direction PP_1 is $\dfrac{1}{r\sin\theta}\cdot D_\phi V$, and in the direction PP_4 is $\dfrac{1}{r}\cdot D_\theta V$; therefore, as the element of volume is made smaller and smaller, our equation approaches the form

$$\sin\theta\cdot D_r(r^2D_rV) + \dfrac{D_\phi^2V}{\sin\theta} + D_\theta(\sin\theta\cdot D_\theta V)$$

$$= -4\pi\rho r^2\sin\theta. \qquad [116]$$

This equation, as well as that for cylinder coördinates, might have been obtained by transformation from the equation in rectangular coördinates.

36. Poisson's Equation in the Integral Form. In [109] X may be regarded as a function of x, y, z, Δy, and Δz, which approaches $D_x V$ as a limit when Δy and Δz are made to approach zero, and it may not be evident that the limit, when Δx, Δy, and Δz are together made to approach zero, of the fraction $\dfrac{\Delta_x X}{\Delta x}$ is D_x^2V. For this reason it is worth while to establish Poisson's Equation by another method.

It is shown in Section 29 that the volume integral of the quantity $-D_x\!\left(\dfrac{\rho}{r}\right)$, taken throughout a certain region, is the surface integral of $\frac{\rho}{r}\cos a$ taken all over the surface which bounds the region. In this proof we might substitute for $\frac{\rho}{r}$ any other function of the three space coördinates which throughout the region is finite, continuous, and single-valued, and state the results in the shape of the following theorem:

If T is any closed surface and U a function of x, y, and z which for every point inside T has a finite, definite value which changes continuously in moving to a neighboring point, then

$$\iiint D_x U \cdot dx\,dy\,dz = -\int U \cos a\,ds, \qquad [117]$$

$$\iiint D_y U \cdot dx\,dy\,dz = -\int U \cos \beta\,ds, \qquad [118]$$

and $$\iiint D_z U \cdot dx\,dy\,dz = -\int U \cos \gamma\,ds, \qquad [119]$$

where a, β, and γ are the angles made by the interior normals at the various points of the surface with the positive direction of the coördinate axes, and where the sinister integrals are to be extended all through the space enclosed by T, and the dexter integrals all over the bounding surface.

If we apply this theorem to an imaginary closed surface which shuts in any attracting mass of density either uniform or variable, and if for U in [117], [118], and [119] we use respectively $D_x V$, $D_y V$, and $D_z V$, and add the resulting equations together, we shall have

$$\iiint (D_x^2 V + D_y^2 V + D_z^2 V)\,dx\,dy\,dz$$

$$= -\int (D_x V \cos a + D_y V \cos \beta + D_z V \cos \gamma)\,ds. \qquad [120]$$

The integral in the second member of this equation is evidently (see [56]) the surface integral of normal attraction taken over our imaginary closed surface, and this by Gauss's Theorem is equal to 4π times the quantity of matter inside the surface, so that

$$\iiint (D_x^2 V + D_y^2 V + D_z^2 V)\, dx\, dy\, dz$$

$$= -4\pi \iiint \rho\, dx\, dy\, dz. \qquad [121]$$

Since this equation is true whatever the form of the closed surface, we must have at every point

$$D_x^2 V + D_y^2 V + D_z^2 V = -4\pi\rho.$$

For if throughout any region $\nabla^2 V$ were greater than $-4\pi\rho$, we might take the boundary of this region as our imaginary surface. In this case every term in the sum whose limit gives the sinister of [121] would be greater than the corresponding term in the dexter, so that the equation would not be true. Similar reasoning shuts out the possibility of $\nabla^2 V$'s being less than $-4\pi\rho$.

37. The Average Value of the Potential Function on a Spherical Surface. If, in a field of force due to a mass m concentrated at a point P, we imagine a spherical surface to be drawn so as to exclude P, the surface integral taken over this surface of the value of the potential function due to m is equal to the area of the surface multiplied by the value of the potential function at the centre of the sphere.

To prove this, let the radius of the sphere be a and the distance $[OP]$ of P from its centre c. Take the centre of the sphere for origin and the line OP for the axis of x. Divide the surface of the sphere into zones by means of a series of planes cutting the axis of x perpendicularly at intervals of Δx. The area of each one of these zones is $2\pi a\, dx$, so that the surface integral of $\dfrac{m}{r}$ is

$$\int_a^{+a} \frac{m\, 2\pi a\, dx}{\sqrt{a^2 + c^2 - 2cx}} = -\left[\frac{2\pi m a \sqrt{a^2 + c^2 - 2cx}}{c} \right]_a^{+a},$$

and the value of this, since the radical represents a positive quantity, is $\dfrac{4\pi a^2 m}{c}$, which proves the proposition.

The surface integral of the potential function taken over the sphere divided by the area of the sphere is often called "the average value of the potential function on the spherical surface."

If we have any distribution of attracting matter, we may divide it into elements, apply the theorem just proved to each of these elements, and, since the potential function due to the whole distribution is the sum of those due to its parts, assert that:

The average value on a spherical surface of the potential function due to any distribution of matter entirely outside the sphere is equal to the value of the potential function at the centre of the sphere.

It follows, from this theorem, that if the potential function is constant within any closed surface S drawn in a region T, which contains no matter, so as to shut in a part of that region, it will have the same value in those parts of T which lie outside S. For, if the values of the potential function at points in empty space just outside S were different from the value inside, it would always be possible to draw a sphere enclosing no matter whose centre should be inside S, and which outside S should include only such points as were all at either higher or lower potential than the space inside S; but in this case the value of the potential function at the centre of the sphere would not be the average of its values over its surface.

The value of the potential function cannot be constant in unlimited empty space surrounding an attracting mass M, for, if it were, we could surround the mass by a surface over which the surface integral of normal attraction would be zero instead of $4\pi M$.

The average value on a spherical surface of the potential function [V], due to any distribution [M] of attracting matter wholly within the surface, is the same as if M were concentrated at the centre O of the space which the surface encloses. For the average values [V_0 and $V_0 + \Delta, V_0$] of V on concentric spherical surfaces of radii r and $r + \Delta r$ may be written

$\dfrac{1}{4\pi r^2}\displaystyle\int V ds$ (or $\dfrac{1}{4\pi}\displaystyle\int V d\omega$, if $d\omega$ is the solid angle of an elementary cone with vertex at O, which intercepts the element ds from the surface of a sphere of radius r), and $\dfrac{1}{4\pi}\displaystyle\int (V+\Delta_r V)d\omega$;

whence

$$\Delta_r V_0 = \frac{1}{4\pi}\int \Delta_r V \cdot d\omega,$$

and

$$D_r V_0 = \frac{1}{4\pi}\int D_r V \cdot d\omega.$$

Now $-\displaystyle\int D_r V \cdot a^2 d\omega$ is the integral of normal attraction taken over the spherical surface, whence, by Gauss's Theorem,

$$D_r V_0 = -\frac{4\pi M}{4\pi r^2}, \quad \text{and} \quad V_0 = \frac{M}{r}+0,$$

since $V_0 = 0$, for $r = \infty$.

38. The Equilibrium of Fluids at Rest under the Action of Given Forces. Elementary principles of Hydrostatics teach us that when an incompressible fluid is at rest under the action of any system of applied forces, the hydrostatic pressure p at the point (x, y, z) must satisfy the differential equation

$$dp = \rho(X dx + Y dy + Z dz), \qquad [122]$$

where X, Y, and Z are the values at that point of the force applied per unit of mass to urge the liquid in directions parallel to the coördinate axes.

For, if we consider an element of the liquid $[\Delta x\, \Delta y\, \Delta z]$ (Fig. 27) whose average density is ρ_0 and whose corner next the origin has the coördinates (x, y, z), and if we denote by p_x the average pressure per unit surface on the face $PP_2 P_4 P_3$, by $p_x + \Delta_x p_x$ the average pressure on the face $P_1 P_5 P_7 P_6$, and by X_0 the average applied force per unit of mass which tends to move the element in a direction parallel to the axis of x, we have, since the element is at rest,

$$p_x \Delta y\, \Delta z + \rho_0 X_0 \Delta x\, \Delta y\, \Delta z = (p_x + \Delta_x p_x)\, \Delta y\, \Delta z,$$

or

$$\rho_0 X_0 = \frac{\Delta_x p_x}{\Delta x}.$$

If the element be made smaller and smaller, the first side of the equation approaches the limit ρX, and the second side the limit $D_x p$, where p is the hydrostatic pressure, equal in all directions, at the point P.

This gives us
$$D_x p = \rho X. \qquad [123]$$

In a similar manner, we may prove that
$$D_y p = \rho Y,$$

and
$$D_z p = \rho Z;$$

whence
$$dp = D_x p\, dx + D_y p\, dy + D_z p\, dz$$
$$= \rho(X\, dx + Y\, dy + Z\, dz).$$

If in any case of a liquid at rest the only external force applied to each particle is the attraction due to some outside mass, or to the other particles of the liquid, or to both together, X, Y, and Z are the partial derivatives with regard to x, y, and z of a single function V, and we may write our general equation in the form
$$dp = \rho(D_x V \cdot dx + D_y V \cdot dy + D_z V \cdot dz) = \rho \cdot dV,$$

whence, if ρ is constant,
$$p = \rho V + \text{const.}, \qquad [124]$$

and the surfaces of equal hydrostatic pressure are also equipotential surfaces.

According to this, the free bounding surfaces of a liquid at rest under the action of gravitation only are equipotential.

EXAMPLES.

1. Prove that a particle cannot be in stable equilibrium under the attraction of any system of masses. [Earnshaw.]

2. Prove that if all the attracting mass lies without an equipotential surface S on which $V = c$, then $V = c$ in all space inside S.

3. Prove that if all the attracting mass lies within an equipotential surface S on which $V = C$, then in all space outside S the value of the potential function lies between C and 0.

4. The source of the Mississippi River is nearer the centre of the earth than the mouth is. What can be inferred from this about the slope of level surfaces on the earth?

5. If in [59] x be made equal to zero, V becomes infinite. How can you reconcile this with what is said in the first part of Section 22?

6. Are all solutions of Laplace's Equation possible values of the potential function in empty space due to distributions of matter? Assume some particular solution of this equation which will serve as the potential function due to a possible distribution and show what this distribution is.

7. If the lines of force which traverse a certain region are parallel, what may be inferred about the intensity of the force within the region?

8. The path of a material particle starting from rest at a point P and moving under the action of the attraction of a given mass M is not in general the line of force due to M which passes through P. Discuss this statement, and consider separately cases where the lines of force are straight and where they are curved.

9. Draw a figure corresponding to Figure 17 for the case of a uniform sphere of unit radius surrounded by a concentric spherical shell of radii 2 and 3 respectively.

10. Draw with the aid of compasses traces of four of the equipotential surfaces due to two homogeneous infinite cylinders of equal density whose axes are parallel and at a distance of 5 inches apart, assuming the radius of one of the cylinders to be 1 inch and that of the other to be 2 inches.

11. Draw with the aid of compasses meridian sections of four of the equipotential surfaces due to two small homogeneous spheres of mass m and $2m$ respectively, whose centres are 4 inches apart. Can equipotential surfaces be drawn so as to lie wholly or partly within one of the spheres? What value of the potential function gives an equipotential surface shaped like the figure 8? Show that the value of the resultant force at the point where this curve crosses itself is zero.

12. A sphere of radius 3 inches and of constant density μ is surrounded by a spherical shell concentric with it of radii 4 inches and 5 inches and of density μr, where r is the distance from the centre. Compute the values of the attraction and of the potential function at all points in space and draw curves to illustrate the fact that V and $D_r V$ are everywhere continuous and that $D_r^2 V$ is discontinuous at certain points.

13. A very long cylinder of radius 4 inches and of constant density μ is surrounded by a cylindrical shell coaxial with it and of radii 6 inches and 8 inches. The density of this shell is inversely proportional to the square of the distance from the axis, and at a point 8 inches from this axis is μ. Use the Theorem of Gauss to find the values of V, $D_r V$, and $D_r^2 V$ at different points on a line perpendicular to the axis of the cylinder at its middle point. If the value of the attraction at a distance of 20 inches from the axis is 10, show how to find μ.

14. Use Dirichlet's value of $D_x V$, given by equation [78], to find the attraction in the direction of the axis of x at points within a spherical shell of radii r_0 and r_1 and of constant density ρ.

15. Are there any other cases except those in which the density of the attracting matter depends only upon the distance from a plane, from an axis, or from a central point, where surfaces of equal force are also equipotential surfaces? Prove your assertion.

16. Prove that if a mass M_1 be divided up into elements, and if each one of these elements be multiplied by the value at one of its own points of the potential function V_2 due to another mass M_2, the limit of the sum of these infinitesimal products will be equal to the limit of the sum extended over M_2 of the product of the masses of its elements by the corresponding values of the potential function due to M_1. That is, show that

$$\int V_2 \cdot dM_1 = \int V_1 \cdot dM_2,$$

where the sinister integral is to embrace all M_1 and the dexter all M_2. [Gauss.]

17. Two uniform straight wires of length l and of masses m_1 and m_2 are parallel to each other and perpendicular to the line joining their middle points, which is of length y_1. Show that the amount of work required to increase the distance between the wires to y_2 by moving one of them parallel to itself is

$$\frac{m_1 m_2}{l^2}\left[y - \sqrt{l^2 + y^2} - l\log\frac{\sqrt{l^2 + y^2} - l}{y}\right]_{y=y_1}^{y=y_2}. \quad \text{[Minchin.]}$$

18. Show that if the earth be supposed spherical and covered with an ocean of small depth, and if the attraction of the particles of water on each other be neglected, the ellipticity of the ocean spheroid will be given by the equation,

$$2e = \frac{\textit{The centrifugal force at the equator}}{g}.$$

19. A spherical shell whose inner radius is r contains a mass m of gas which obeys the Law of Boyle and Mariotte. Find the law of density of the gas, the total normal pressure on the inside of the containing vessel, and the pressure at the centre.

20. If the earth were melted into a sphere of homogeneous liquid, what would be the pressure at the centre in tons per square foot? If this molten sphere instead of being homogeneous had a surface density of 2.4 and an average density of 5.6, what would be the pressure at the centre on the supposition that the density increased proportionately to the depth?

21. A solid sphere of attracting matter of mass m and of radius r is surrounded by a given mass M of gas which obeys the Law of Boyle and Mariotte. If the whole is removed from the attraction of all other matter, find the law of density of the gas and the pressure on the outside of the sphere.

22. The potential function within a closed surface S due to matter wholly outside the surface has for extreme values the extreme values upon S.

23. If the potential functions V and V' due to two systems of matter without a closed surface have the same values at all points on the surface, they will be equal throughout the space enclosed by the surface.

24. The potential function outside of a closed surface due to matter wholly within the surface has for its extreme values two of the following three quantities : zero and the extreme values upon the surface.

25. Prove that if R is the distance from the origin of coördinates to the point P, and if V_p is the value at P of the potential function of any system of attracting masses within a finite distance of the origin, the limit as R is made infinite of $V_p \cdot P$ is equal to M, the whole quantity of attracting matter.

CHAPTER III.

THE POTENTIAL FUNCTION IN THE CASE OF REPULSION.

39. Repulsion, according to the Law of Nature.   Certain physical phenomena teach us that bodies may acquire, by electrification or otherwise, the property of repelling each other, and that the resulting force of repulsion between two bodies is often much greater than the force of attraction which, according to the Law of Gravitation, every body has for every other body.

Experiment shows that almost every such case of repulsion, however it may be explained physically, can be quantitatively accounted for by assuming the existence of some distribution of a kind of " matter," every particle of which is supposed to repel every other particle of the same sort according to the " Law of Nature," that is, roughly stated, with a force directly proportional to the product of the quantities of matter in the particles, and inversely proportional to the square of the distance between their centres.

In this chapter we shall assume, for the sake of argument, that such matter exists, and proceed to discuss the effects of different distributions of it.   Since the law of repulsion which we have assumed is, with the exception of the opposite directions of the forces, mathematically identical with the law which governs the attraction of gravitation between particles of ponderable matter, we shall find that, by the occasional introduction of a change of sign, all the formulas which we have proved to be true for cases of attraction due to gravitation can be made useful in treating corresponding problems in repulsion.

40. Force at Any Point due to a Given Distribution of Repelling Matter. Two equal quantities of repelling matter concentrated at points at the unit distance apart are called "unit quantities" when they are such as to make the force of repulsion between them the unit force.

If the ratio of the quantity of repelling matter within a small closed surface supposed drawn about a point P, to the volume of the space enclosed by the surface, approaches the limit ρ when the surface (always enclosing P) is supposed to be made smaller and smaller, ρ is called the "density" of the repelling matter at P.

In order to find the magnitude at any point P of the force due to any given distribution of repelling matter, we may suppose the space occupied by this matter to be divided up into small elements, and compute an approximate value of this force on the assumption that each element repels a unit quantity of matter concentrated at P with a force equal to the quantity of matter in the element divided by the square of the distance between P and one of the points of the element. The limit approached by this approximate value as the size of the elements is diminished indefinitely is the value required.

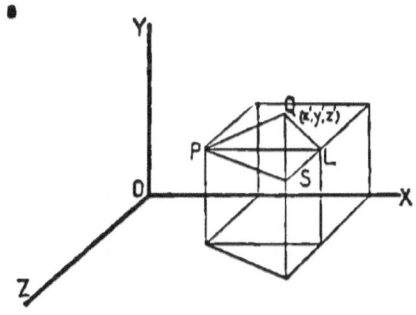

Fig. 30.

Let Q (Fig. 30), whose coördinates are x', y', z', be the corner next the origin of an element of the distribution. Let ρ be the density at Q and $\Delta x'\Delta y'\Delta z'$ the volume of the element; then the force at P due to the matter in the element is approximately equivalent to a force of magnitude $\dfrac{\rho\,\Delta x'\,\Delta y'\,\Delta z'}{PQ^2}$ acting in

the direction QP, or a force of magnitude $-\dfrac{\rho\,\Delta x'\,\Delta y'\,\Delta z'}{PQ^2}$ acting

in the direction PQ. If the coördinates of P are x, y, z, the component of this force in the direction of the positive axis of x

is $\dfrac{-\rho\,\Delta x'\,\Delta y'\,\Delta z'\,(x'-x)}{[(x'-x)^2+(y'-y)^2+(z'-z)^2]^{\frac{3}{2}}}$, and the force at P parallel

to the axis of x due to the whole distribution of repelling matter is

$$X=-\iiint\frac{\rho(x'-x)\,dx'dy'dz'}{[(x'-x)^2+(y'-y)^2+(z'-z)^2]^{\frac{3}{2}}}, \qquad [125_\text{A}]$$

where the triple integration is to be extended over the whole space filled with the repelling matter. For the components of the force at P parallel to the other axes we have, similarly,

$$Y=-\iiint\frac{\rho(y'-y)\,dx'dy'dz'}{[(x'-x)^2+(y'-y)^2+(z'-z)^2]^{\frac{3}{2}}}, \qquad [125_\text{B}]$$

and

$$Z=-\iiint\frac{\rho(z'-z)\,dx'dy'dz'}{[(x'-x)^2+(y'-y)^2+(z'-z)^2]^{\frac{3}{2}}}. \qquad [125_\text{C}]$$

If we denote by V the function

$$\iiint\frac{\rho\,dx'dy'dz'}{[(x'-x)^2+(y'-y)^2+(z'-z)^2]^{\frac{1}{2}}}, \qquad [126]$$

which, together with its first derivatives, is everywhere finite and continuous, as we have shown in the last chapter, it is easy to see that

$$X=-D_xV, \quad Y=-D_yV, \quad Z=-D_zV, \qquad [127]$$

$$R=\sqrt{(D_xV)^2+(D_yV)^2+(D_zV)^2}, \qquad [128]$$

and that the direction-cosines of the line of action of the resultant force at P are

$$-\frac{D_xV}{R}, \quad -\frac{D_yV}{R}, \quad \text{and} \quad -\frac{D_zV}{R}. \qquad [129]$$

It follows from this (see Section 21) that the component in any direction of the force at a point P due to any distribution M of repelling matter is minus the value at P of the partial derivative of the function V taken in that direction.

The function V goes by the name of the Newtonian potential function whether we are dealing with attracting or repelling matter.

In the case of repelling matter, it is evident that the resultant force on a particle of the matter at any point tends to drive that particle in a direction which leads to points at which the potential function has a lower value, whereas in the case of gravitation a particle of ponderable matter at any point tends to move in a direction along which the potential function increases.

41. The Potential Function as a Measure of Work. It is easy to show by a method like that of Article 27 that the amount of work required to move a unit quantity of repelling matter, concentrated at a point, from P_1 to P_2, in face of the force due to any distribution M of the same kind of matter, is $V_2 - V_1$, where V_1 and V_2 are the values at P_1 and P_2 respectively of the potential function due to M. The farther P_1 is from the given distribution, the smaller is V_1, and the less does $V_2 - V_1$ differ from V_2. In fact, the value of the potential function at the point P_2, wherever it may be, measures the work which would be required to move the unit quantity of matter by any path from "infinity" to P_2.

42. Gauss's Theorem in the Case of Repelling Matter. If a quantity m of repelling matter is concentrated at a point within a closed oval surface, the resultant force due to m at any point on the surface acts toward the outside of the surface instead of towards the inside, as in the case of attracting matter.

Keeping this in mind, we may repeat the reasoning of Article 31, using repelling matter instead of attracting matter, and substituting all through the work the exterior normal for the interior normal, and in this way prove that:

If there be any distribution of repelling matter partly within and partly without a closed surface T, and if M be the whole quantity of this matter enclosed by T, and M' the quantity outside T, the surface integral over T of the component in the direction of the *exterior* normal of the force due to both M and M' is equal to $4\pi M$. If V be the potential function due to M and M', we have

$$\int D_n V \cdot ds = 4\pi M.$$

43. Poisson's Equation in the Case of Repelling Matter. If we apply the theorem of the last article to the surface of a volume element cut out of space containing repelling matter, and use the notation of Article 35, we shall find that in the case of rectangular coördinates the surface integral, taken over the element, of the component in the direction of the exterior normal is

$$\Delta x \Delta y \Delta z \left[\frac{\Delta_x X}{\Delta x} + \frac{\Delta_y Y}{\Delta y} + \frac{\Delta_z Z}{\Delta z} \right] = 4\pi \rho_0 \cdot \Delta x \Delta y \Delta z, \quad [130]$$

where X is the average component in the positive direction of the axis of x of the force on the elementary surface $\Delta y \Delta z$, and where Y and Z have similar meanings. It is evident that if the element be made smaller and smaller, X, Y, and Z will approach as limits the components parallel to the coördinate axes of the force at P. These components are $-D_x V$, $-D_y V$, and $-D_z V$; so that if we divide [130] by $\Delta x \Delta y \Delta z$ and then decrease indefinitely the dimensions of the element, we shall arrive at the equation

$$\nabla^2 V = -4\pi\rho. \quad [131]$$

By using successively cylinder coördinates and spherical coördinates we may prove the equations

$$\frac{1}{r} D_r(r D_r V) + \frac{1}{r^2} D_\theta^2 V + D_z^2 V = -4\pi\rho, \quad [132]$$

and $$\sin\theta \cdot D_r(r^2 D_r V) + \frac{D_\phi^2 V}{\sin\theta} + D_\theta(\sin\theta \cdot D_\theta V)$$

$$= -4\pi\rho r^2 \sin\theta, \quad [133]$$

so that Poisson's Equation holds whether we are dealing with attracting or repelling matter.

44. Coexistence of Two Kinds of Active Matter. Certain physical phenomena may be most conveniently treated mathematically by assuming the coexistence of two kinds of "matter" such that any quantity of either kind repels all other matter of the same kind according to the Law of Nature, and attracts all matter of the other kind according to the same law.

Two quantities of such matter may be considered equal if, when placed in the same position in a field of force, they are subjected to resultant forces which are equal in intensity and which have the same line of action. The two quantities of matter are of the same kind if the direction of the resultant forces is the same in the two cases, but of different kinds if the directions are opposed. The unit quantity of matter is that quantity which concentrated at a point would repel with the unit force an equal quantity of the same kind concentrated at a point at the unit distance from the first point.

It is evident from Articles 2, 14, and 40 that m units of one of these kinds of matter, if concentrated at a point (x, y, z) and exposed to the action of $m_1, m_2, m_3, \ldots m_k$ units of the same kind of matter concentrated respectively at the points (x_1, y_1, z_1), (x_2, y_2, z_2), (x_3, y_3, z_3), $\ldots (x_k, y_k, z_k)$, and of $m_{k+1}, m_{k+2} \ldots m_n$ units of the other kind of matter concentrated respectively at the points $(x_{k+1}, y_{k+1}, z_{k+1})$, $(x_{k+2}, y_{k+2}, z_{k+2})$, $\ldots (x_n, y_n, z_n)$, will be urged in the direction parallel to the positive axis of x with the force

$$X = -m \sum_{i=1}^{i=k} \frac{m_i (x_i - x)}{r_i^3} + m \sum_{i=k+1}^{i=n} \frac{m_i (x_i - x)}{r_i^3}, \qquad [134]$$

where r_i is the distance between the points (x, y, z) and (x_i, y_i, z_i).

If we agree to distinguish the two kinds of matter from each other by calling one kind "positive" and the other kind "negative," it is easy to see that if every m which belongs to positive matter be given the plus sign and every m which belongs to negative matter the minus sign, we may write the last equation in the form

$$X = -m \sum_{i=1}^{i=n} \frac{m_i(x_i - x)}{r_i^3}. \qquad [135]$$

The result obtained by making m in [135] equal to unity is called the force at the point (x, y, z).

In general, m units of either kind of matter concentrated at the point (x, y, z), and exposed to the action of any continuous distribution of matter, will be urged in the positive direction of the axis of x by the force

$$X = -m \iiint \frac{\rho(x' - x)\,dx'dy'dz'}{[(x'-x)^2 + (y'-y)^2 + (z'-z)^2]^{\frac{3}{2}}}; \qquad [136]$$

in this expression, ρ, the density at (x', y', z'), is to be taken positive or negative according as the matter at the point is positive or negative: m is to have the sign belonging to the matter at the point (x, y, z): and the limits of integration are to be chosen so as to include all the matter which acts on m.

With the same understanding about the signs of m and of ρ, it is clear that the force which urges in any direction s, m units of matter concentrated at the point (x, y, z) is equal to $-m \cdot D_s V$, where V is the everywhere finite, continuous, and single-valued function

$$\iiint \frac{\rho(x' - x)\,dx'\,dy'\,dz'}{[(x'-x)^2 + (y'-y)^2 + (z'-z)^2]^{\frac{1}{2}}};$$

and that mV measures the amount of work required to bring up from "infinity" by any path to its present position the m units of matter now at the point (x, y, z).

If we call the resultant force which would act on a unit of positive matter concentrated at the point P "the force at P," it is clear that if any closed surface T be drawn in a field of force due to any distribution of positive and negative matter so as to include a quantity of this matter algebraically equal to Q, the surface integral taken over T of the component in the direction of the exterior normal of the force at the different points of the surface is equal to $4\pi Q$.

It will be found, indeed, that all the equations and theorems given earlier in this chapter for the case of one kind of repelling matter may be used unchanged for the case where positive and negative matter coexist, if we only give to ρ and m their proper signs.

It is to be noticed that Poisson's Equation is applicable whether we are dealing with attracting matter or repelling matter, or positive and negative matter existing together.

EXAMPLES.

1. Show that the extreme values of the potential function outside a closed surface S, due to a quantity of matter algebraically equal to zero within the surface, are its extreme values on S.

2. Show that if the potential function due to a quantity of matter algebraically equal to zero and shut in by a closed surface S has a constant value all over the surface, then this constant value must be zero.

CHAPTER IV.

SURFACE DISTRIBUTIONS.—GREEN'S THEOREM.

45. Force due to a Closed Shell of Repelling Matter. If a quantity of very finely-divided repelling matter be enclosed in a box of any shape made of indifferent material, it is evident from [127] and from the principles of Section 38 that if the volume of the box is greater than the space occupied by the repelling matter, the latter will arrange itself so that its free surface will be equipotential with regard to all the active matter in existence, taking into account any there may be outside the box as well as that inside. It is easy to see, moreover, that we shall have a shell of matter lining the box and enclosing an empty space in the middle.

That any such distribution as that indicated in the subjoined diagram is impossible follows immediately from the reasoning of Section 37. For *ABC* and *DEF* are parts of the same

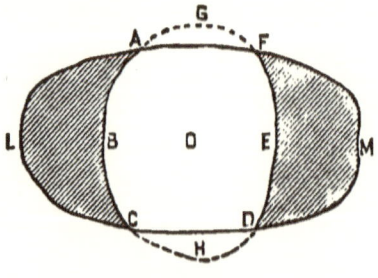

Fig. 31.

equipotential free surface of the matter. If we complete this surface by the parts indicated by the dotted lines, we shall enclose a space void of matter and having therefore throughout a value of the potential function equal to that on the bounding surface. But in this case all points which can be reached from O by paths which do not cut the repelling matter must be at the same potential as O, and this evidently includes all space not actually occupied by the repelling matter; which is absurd.

Let us consider, then (see Fig. 32), a closed shell of repelling matter whose inner surface is equipotential, so that at every point of the cavity which the shell shuts in, the resultant force, due to the matter of which the shell is composed and to any outside matter there may be, is zero.

Let us take a small portion ω of the bounding surface of the cavity as the base of a tube of force which shall intercept an

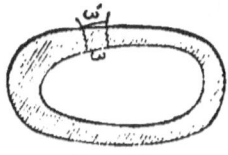

Fig. 32.

area ω' on an equipotential surface which cuts it just outside the outer surface of the shell, and let us apply Gauss's Theorem to the box enclosed by ω, ω', and the tube of force. If F' is the average value of the resultant force on ω', the only part of the surface of the box which yields anything to the surface integral of normal force, we have

$$F'\omega' = 4\pi m,$$

where m is the quantity of matter within the box. If we multiply and divide by ω, this equation may be written

$$F' = \frac{4\pi m}{\omega} \cdot \frac{\omega}{\omega'}. \qquad [137]$$

If ω be made smaller and smaller, so as always to include a given point A, ω' as it approaches zero will always include a point B on the line of force drawn through A, and F' will approach the value F of the resultant force at B.

The shell may be regarded as a thick layer spread upon the

inner surface, and in this case the limit of $\frac{m}{\omega}$ may be considered the value at A of the rate at which the matter is spread upon the surface. If we denote this limit by σ, we shall have

$$F = 4\pi\sigma \cdot \underset{\omega \doteq 0}{\text{limit}} \left(\frac{\omega}{\omega'}\right). \qquad\qquad [138]$$

If B be taken just outside the shell, and if the latter be very thin, $\underset{\omega \doteq 0}{\text{limit}} \left(\frac{\omega}{\omega'}\right)$ evidently differs little from unity; and we see that the resultant force at a point just outside the outer surface of a shell of matter, whose inner surface is equipotential, becomes more and more nearly equal to 4π times the quantity of matter per unit of surface in the distribution at that point as the shell becomes thinner and thinner.

The reader may find out for himself, if he pleases, whether or not the line of action of the resultant force at a point just outside such a shell as we have been considering is normal to the shell.

It is to be carefully noticed that the inner surface of a closed shell need not be equipotential unless the matter composing the shell is finely divided and free to arrange itself at will.

When the shell is thin, and we regard it as formed of matter spread upon its inner surface, σ is called the "surface density" of the distribution, and its value at any point of the inner surface of the shell may be regarded as a measure of the amount of matter which must be spread upon a unit of surface if it is to be uniformly covered with a layer of thickness equal to that of the shell at the point in question.

46. Surface Distributions. It often becomes necessary in the mathematical treatment of physical problems, on the assumption of the existence of a kind of repelling matter or agent, to imagine a finite quantity of this agent *condensed on a surface* in a layer so thin that for practical purposes we may leave the thickness out of account. If a shell like that considered in the last section could be made thinner and thinner by compression while the quantity of matter in it remained unchanged, the volume density (ρ) of the shell would grow larger and larger without limit, and σ would remain finite. A distribution like this, which is considered to have *no* thickness, is called a surface distribution.

The value at a point P of the potential function due to a superficial distribution of surface density σ is the surface integral, taken over the distribution, of $\frac{\sigma}{r}$, where r is the distance from P.

It is evident that as long as P does not lie exactly in the distribution, the potential function and its derivatives are always finite and continuous, and the force at any point in any direction may be found by differentiating the potential function partially with regard to that direction.

If ρ were infinite, the reasoning of Article 22 would no longer apply to points actually in the active matter, and it is worth our while to prove that in the case of a surface distribution where σ is everywhere finite, the value at a point P of the potential function due to the distribution remains finite, as

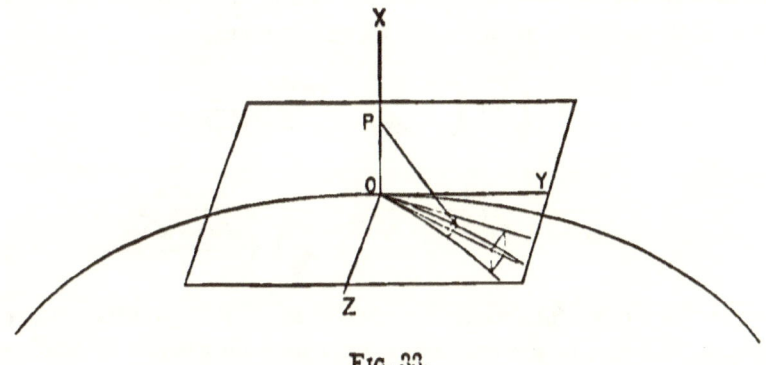

Fig. 33.

P is made to move normally through the surface at a point of finite curvature.

To show this, take the point O (Fig. 33), where P is to cut the surface, as origin, and the normal to the surface at O as the axis of x, so that the other coördinate axes shall lie in the tangent plane.

If the curvature in the neighborhood of O is finite, it will be possible to draw on the surface about O a closed line such that for every point of the surface within this line the normal will make an acute angle with the axis of x.

For convenience we will draw the closed line of such a shape that its projection on the tangent plane shall be a circle whose centre is at O and whose radius is U, and we will cut the area shut in by this line into elements of such shape that their projections upon the tangent plane shall divide the circle just mentioned into elements bounded by concentric circumferences drawn at radial intervals of Δu, and by radii drawn at angular distances of $\Delta\phi$.

If $x, 0, 0$ are the coördinates of the point P, x', y', z' those of a point of one of the elements of the area shut in by the closed line, and a the angle which the normal to the surface at this point makes with the axis of x, the size of the surface element is approximately $\dfrac{u\,\Delta u\,\Delta\phi}{\cos a}$, where $u^2 = z'^2 + y'^2$, and the value at P of the potential function due to that part of the surface distribution shut in by the closed line is

$$V_1 = \int_0^{2\pi} d\phi \int_0^U \frac{\sigma u\,du}{\cos a \sqrt{(x-x')^2 + u^2}}. \qquad [139]$$

The quantity

$$\frac{\sigma u}{\cos a \sqrt{(x-x')^2 + u^2}} = \frac{\sigma \sec a}{\sqrt{1 + \left(\dfrac{x-x'}{u}\right)^2}}$$

is always finite, for, whatever the value of the quantity under the radical sign in the last expression may be when x, x', and u are all zero, it cannot be less than unity, and therefore V_1 must be finite even when P moves down the axis of x to the surface itself.

If V and V_2 are the values at P of the potential functions due respectively to all the existing acting matter and to that part of this matter not lying on the portion of the surface shut in by our closed line, we have $V = V_1 + V_2$, and, since P is a point outside the matter which gives rise to V_2, the latter is finite; so that V is finite.

The reader who wishes to study the properties of the derivatives of the potential function, and their relations to the force components at points actually in a surface distribution, will find the whole subject treated in the first part of Riemann's *Schwere, Electricität und Magnetismus.*

Using the notation of this section, it is easy to write down definite integrals which represent the values of the potential function at two points on the same normal, one on one side of a superficial distribution, and at a distance a from it, and the other on the other side at a like distance, and to show that the difference between these integrals may be made as small as we like by choosing a small enough. This shows that the value of the potential function at a point P changes continuously, as P moves normally through a surface distribution of finite superficial density. If matter could be concentrated upon a geometric line, so that there should be a finite quantity of matter on the unit of length of the line, or if a finite quantity of matter could be really concentrated at a point, the resulting potential function would be infinite on the line itself, and at the point.

47. The Normal Force at Any Point of a Surface Distribution. In the case of a strictly superficial distribution on a closed surface where the repelling matter is free to arrange itself at will, the inner surface of the matter (and hence the outer surface, which is coincident with it) is equipotential, and the resultant force at a point B just outside the distribution is normal to the surface and numerically equal to 4π times the surface density at B. This shows that the derivative of the potential function in the direction of the normal to the surface has values on opposite sides of the surface differing by $4\pi\sigma$, and at the surface itself cannot be said to have any definite value.

It is easy, however, to find the force with which the repelling matter composing a superficial distribution is urged outwards. For, take a small element ω of the surface as the base of a tube of force, and apply Gauss's Theorem to a box shut in by the surface of distribution, the tube of force, and a portion ω' of an equipotential surface drawn just outside the distribution. Let F and F' be the average forces at the points of ω and ω' respectively, then the surface integral of normal forces taken over the box is $F'\omega' - F\omega$, and this, since the only active matter is concentrated on the surface of the box (see Section 31), is equal to $2\pi\sigma_0\omega$, where σ_0 is the average surface density at the points of the element ω. This gives us

$$F = F'\,\frac{\omega'}{\omega} - 2\pi\sigma_0.$$

Now let the equipotential surface of which ω' is a part be drawn nearer and nearer the distribution; then

$$\lim\frac{\omega'}{\omega} = 1, \quad \lim F' = 4\pi\sigma_0, \quad \text{and} \quad F = 2\pi\sigma_0.$$

F is the average force which would tend to move a unit quantity of repelling matter concentrated successively at the different points of ω in the direction of the exterior normal, but the actual distribution on ω is $\omega\sigma_0$, so that this matter presses on the medium which prevents it from escaping with the force $2\pi\sigma_0^2\omega$; and, in general, the pressure exerted on the resisting medium which surrounds a surface distribution of repelling matter is at any point $2\pi\sigma^2$ per unit of surface, where σ is the surface density of the distribution at the point in question.

We may imagine a superficial distribution of matter which is fixed, instead of being free to arrange itself at will. In this case the surface of the matter will not be in general equipotential, but, if we apply Gauss's Theorem to a box shut in by a slender tube of force traversing the distribution, and by two surfaces drawn parallel to the distribution and close to it, one on one side and one on the other, we may prove that the normal component of the force at a point just outside the distribution differs by $4\pi\sigma$ from the normal component, in the same sense, of the force at a point just inside the distribution on the line of force which passes through the first point.

48. Green's Theorem. Before proving a very general theorem due to Green,* of which what we have called Gauss's Theorem is a special case, we will show that if T is any closed surface and U a function of x, y, and z, which for every point inside T is finite, continuous, and single-valued,

$$\int\int\int D_x U \cdot dx\,dy\,dz = \int U \cdot D_n x \cdot ds, \qquad [140]$$

where the first integral is to include all the space shut in by T, and the second is to be taken over the whole surface, and where $D_n x$ represents the partial derivative of x taken in the direction of the *exterior* normal.

To prove this, choose the coördinate axes so that T shall lie in the first octant, and divide the space inside the contour of the

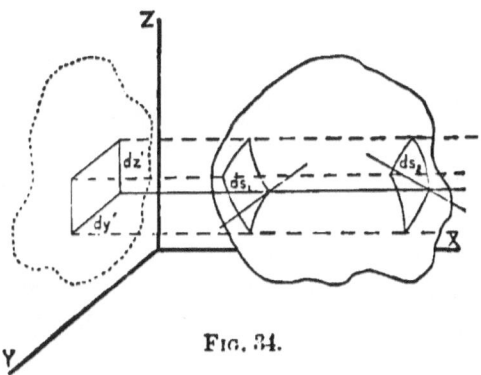

Fig. 34.

projection of T on the plane yz into elements of size $dy\,dz$. On each of these elements erect a right prism cutting T twice or some other even number of times. Let us call the values of U at the successive points where the edge nearest the axis of x of

* George Green, *An Essay on the Application of Mathematical Analysis to the Theories of Electricity and Magnetism.* Nottingham, 1828.

any one of these prisms cuts T, U_1, U_2, U_3, ... U_{2n} respectively ; the angles which this edge makes with exterior normals drawn to T at these points, a_1, a_2, a_3, ... a_{2n} ; and the elements which the prism cuts from the surface T, ds_1, ds_2, ds_3, ... ds_{2n}. It is evident that wherever a line perpendicular to the plane yz cuts *into* T, the corresponding value of a is obtuse and its cosine negative, but wherever such a line cuts *out* of T, the corresponding value of a is acute and its cosine positive.

Keeping this in mind, we shall see that although the base of a prism is the common projection of all the elements which it cuts from T, and in absolute value is approximately equal to any one of these multiplied by the corresponding value of cos a, yet, since $dx\,dy$, ds_1, ds_2, etc., are all positive areas and some of the cosines are negative, we must write, if we take account of signs,

$$dy\,dz = -ds_1 \cos a_1 = +ds_2 \cos a_2 = -ds_3 \cos a_3 = \cdots.$$

If the indicated integration with regard to x in the left-hand member of [140] be performed and the proper limits introduced, we shall have

$$\iiint D_x U\,dx\,dy\,dz = \iint dy\,dz[-U_1 + U_2 - U_3 + U_4 - \cdots], \quad [141]$$

where the double sign of integration directs us to form a quantity corresponding to that in brackets for every prism which cuts T, to multiply this by the area of the base of the prism, and to find the limit of the sum of all the results as the bases of the prisms are made smaller and smaller.

Since we may substitute for $dy\,dz$ any one of its approximate values given above, we may write the quantity within the brackets

$$U_1 \cos a_1\,ds_1 + U_2 \cos a_2\,ds_2 + U_3 \cos a_3\,ds_3 + \cdots,$$

and this shows that the double integral is equivalent to the surface integral, taken over the whole of T, of $U \cos a$, whence we may write

$$\iiint D_x U \cdot dx\,dy\,dz = \int U \cos a\,ds, \quad [142]$$

where the first integral is to be taken all through the space shut in by T, and the second over the whole surface.

Let $P(x, y, z)$ be any point of T, a, β, and γ the angles which the exterior normal drawn to P at P makes with the coördinate axes, and P' a point on this normal at a distance Δn from P. The coördinates of P' are

$$x + \Delta n \cdot \cos a, \quad y + \Delta n \cdot \cos \beta, \quad z + \Delta n \cdot \cos \gamma,$$

and if $W = f(x, y, z)$ be any continuous function of the space coördinates,

$$W_P = f(x, y, z),$$

$$W_{P'} = f(x + \Delta n \cos a, \ y + \Delta n \cos \beta, \ z + \Delta n \cos \gamma)$$
$$= f(x, y, z) + \Delta n \cos a \cdot D_x f + \Delta n \cos \beta \cdot D_y f$$
$$+ \Delta n \cos \gamma \, D_z f + (\Delta n)^2 Q,$$

and

$$\frac{W_{P'} - W_P}{PP'} = \cos a \cdot D_x f + \cos \beta \cdot D_y f + \cos \gamma \cdot D_z f + \Delta n \cdot Q.$$

whence

$$\lim \frac{W_{P'} - W_P}{PP'} = D_n W_P = \cos a \, D_x f + \cos \beta \, D_y f + \cos \gamma \, D_z f. \quad [143]$$

If, as a special case, $W = x$, we have $D_n x = \cos a$; so that [142] may be written

$$\iiint D_x U \cdot dx \, dy \, dz = \int U D_n x \cdot ds, \qquad [144]$$

which we were to prove.*

Green's Theorem, which follows very easily from this result, may be stated in the following form :

If U and V are any two functions of the space coördinates which together with their first derivatives with respect to these coördinates are finite, continuous, and single-valued throughout the space shut in by any closed surface T, then, if n refers to an exterior normal,

---

* This theorem has been virtually proved already in Sections 29 and 38.

$$\iiint (D_xU \cdot D_xV + D_yU \cdot D_yV + D_zU \cdot D_zV)dx\,dy\,dz$$

$$= \int U \cdot D_nV \cdot ds - \iiint U \cdot \nabla^2 V \cdot dx\,dy\,dz \qquad [145]$$

$$= \int V \cdot D_nU \cdot ds - \iiint V \cdot \nabla^2 U \cdot dx\,dy\,dz, \qquad [146]$$

where the triple integrals include all the space within T and the single integrals include the whole surface.

Since $\qquad D_zU \cdot D_zV = D_z(U \cdot D_zV) - U \cdot D_z^2V,$

we have $\qquad \iiint D_zU \cdot D_zV \cdot dx\,dy\,dz$

$$= \iiint D_z(U \cdot D_zV)dx\,dy\,dz - \iiint U \cdot D_z^2V \cdot dx\,dy\,dz\;;$$

but, from [144],

$$\iiint D_z(U \cdot D_zV)dx\,dy\,dz = \int U \cdot D_zV \cdot D_nx \cdot ds,$$

whence $\qquad \iiint (D_zU \cdot D_zV)dx\,dy\,dz$

$$= \int U \cdot D_zV \cdot D_nx \cdot ds - \iiint U \cdot D_z^2V \cdot dx\,dy\,dz. \quad [147]$$

If we form the two corresponding equations for the derivatives with regard to y and z, and add the three together, we shall obtain an expression which, by the use of [143], reduces immediately to [145]. Considerations of symmetry give [146].

If we subtract [146] from [145], we get

$$\iiint (U \cdot \nabla^2 V - V \cdot \nabla^2 U)dx\,dy\,dz$$

$$= \int (U \cdot D_nV - V \cdot D_nU)ds. \qquad [148]$$

In applying Green's Theorem to such spaces as those marked T_0 in the adjoining diagrams, it is to be noticed that the walls of the cavities, marked S' and S'', as well as the surfaces, marked S, form parts of the boundaries of the spaces, and that the surface integrals, which the theorem declares must be taken

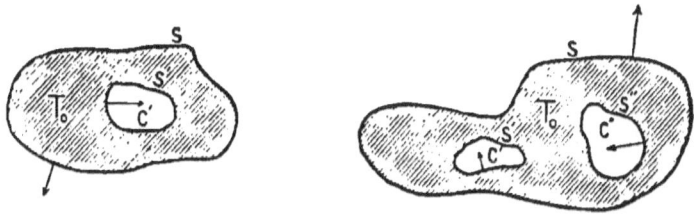

FIG. 35.

over the whole boundaries of the spaces, are to be extended over S' and S'' as well as over S. We must remember, however, that an exterior normal to T_0 at S' points *into* the cavity C'.

49. Special Cases under Green's Theorem. I. If in [148] V be the potential function due to any distribution either of repelling matter or of positive and negative matter existing together, whether this matter is within or without T, and if $U = 1$, we have

$$\nabla^2 V = - 4 \pi \rho,$$

and
$$4 \pi \iiint \rho \, dx \, dy \, dz = \int [-D_n V] \, ds. \qquad [149]$$

The triple integral on the left-hand side of the equation is the whole amount of matter (algebraically considered, where we have both positive and negative matter) within T, and the dexter is the surface integral taken over T of the force in the direction of the exterior normal; so that [149] expresses Gauss's Theorem.

II. If in [145] we make U equal to V, and let this represent as before the potential function due to any distribution of actual matter within or without T, we shall have

$$\iiint R^2 dx \, dy \, dz = \int V \cdot D_n V \, ds + 4 \pi \iiint \rho V \, dx \, dy \, dz, \qquad [150]$$

where R is the resultant force at the point (x, y, z).

III. If in [145] we make $U = V = u$, any function which within the closed surface T satisfies the equation $\nabla^2 u = 0$, we shall have

$$\int\int\int [(D_x u)^2 + (D_y u)^2 + (D_z u)^2]\, dx\, dy\, dz = \int u \cdot D_n u \cdot ds. \quad [151]$$

IV. If in [148] V is the potential function due to two distributions of active matter, M_1 inside the closed surface T and M_2 outside it, and if $U = \dfrac{1}{r}$ where r is the distance of the point (x, y, z) from a fixed point O, we must consider separately the two cases where O is respectively without T and within T.

A. If O is without T, $\nabla^2 \left(\dfrac{1}{r}\right) = 0$ for points within the surface. Also, $\nabla^2 V = -4\pi\rho$, so that

$$\int \frac{D_n V}{r}\, ds - \int V \cdot D_n\left(\frac{1}{r}\right) ds = -4\pi \int\int\int \frac{\rho}{r}\, dx\, dy\, dz.$$

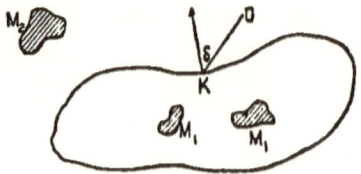

FIG. 36.

The triple integral is evidently equal to the value at the point O of the potential function due to M_1 alone. If we call this V_1, and notice (see [143]) that $D_n r$ at any point of T is the cosine of the angle δ between r and the exterior normal to T, we have

$$\int \frac{D_n V}{r}\, ds - \int \frac{V \cos\delta}{r^2}\, ds = -4\pi V_1. \quad [152]$$

If T is a surface equipotential with respect to the joint action of M_1 and M_2, and if we denote by V_s the constant value of V on T, we have

$$\int \frac{D_n V}{r}\, ds - V_s \int \frac{\cos\delta}{r^2}\, ds = -4\pi V_1,$$

and it is easy to show, by the reasoning used in Section 31, that $\int \frac{\cos \delta}{r^2} ds = 0$, whence

$$V_1 = -\frac{1}{4\pi} \int \frac{D_n V}{r} ds. \qquad [153]$$

B. If O is a point inside T, whether or not it is within M_1, and if T is equipotential with respect to the action of M_1 and M_2, we will surround O by a small spherical surface s' of radius r', and apply [148] to the space lying inside T and without the spherical surface. In doing so, it is to be noticed that s' forms part of the boundary of the region we are dealing with, and that an exterior normal to the region at s' will be an interior normal of the sphere.

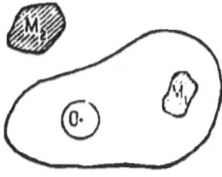

Fig. 37.

Since for all points of the region we are considering $\nabla^2 \left(\frac{1}{r} \right) = 0$, we have

$$\int \frac{D_n V}{r} ds - \int \frac{D_{r'} V}{r'} ds' - V \int D_n \left(\frac{1}{r} \right) ds + \int V D_{r'} \left(\frac{1}{r'} \right) ds'$$
$$= -4\pi \iiint \frac{\rho}{r} dx\,dy\,dz; \qquad [154]$$

or, since $ds' = r'^2 d\omega'$, where $d\omega'$ is the area which the elementary cone whose base is ds' and vertex O intercepts on the sphere of unit radius drawn about O,

$$\int \frac{D_n V}{r} ds + V \int \frac{\cos \delta}{r^2} ds - r' \int D_{r'} V \cdot d\omega' - \int V d\omega'$$
$$= -4\pi \iiint \frac{\rho}{r} dx\,dy\,dz. \qquad [155]$$

It is easily proved, by the reasoning of Section 31, that

$$\int \frac{\cos \delta}{r^2}\, ds = 4\pi,$$

and it is clear that if r' be made smaller and smaller, the third integral of [155] approaches the limit zero. If V' is the average value of V on the surface s',

$$\int V' d\omega = V' \int d\omega = V' 4\pi ;$$

and as r' is made smaller and smaller, this approaches the value $4\pi V_0$, where V_0 is the value of V at O. The value, when r' is zero, of the triple integral in [155] is evidently V_1, and we have

$$\int \frac{D_n V}{r}\, ds + 4\pi V_s - 4\pi V_0 = -4\pi V_1. \qquad [156]$$

If V_2 is the value at O of the potential function due to M_2 alone, $V_0 = V_1 + V_2$, so that [156] may be written in the form

$$V_s - V_2 = -\frac{1}{4\pi} \int \frac{D_n V}{r}\, ds. \qquad [157]$$

If T is not equipotential with respect to the action of M_1 and M_2, we have

$$4\pi V_2 = \int \frac{D_n V}{r}\, ds - \int V D_n \left(\frac{1}{r} \right) ds. \qquad [158]$$

V. If in [148] we make $U = \frac{1}{r}$, where r is the distance of the point (x, y, z) from a fixed point O, and if $V = v$, a function which within the closed surface T satisfies the equation $\nabla'' v = 0$, we shall have

$$4\pi v = \int v D_n \left(\frac{1}{r} \right) ds - \int \frac{D_n v}{r}\, ds, \qquad [159]$$

if O is within T, and

$$\int \frac{D_n v}{r}\, ds = \int v \cdot D_n \left(\frac{1}{r} \right) ds, \qquad [160]$$

if O is outside T.

50. The Surface Distributions Equivalent to Certain Volume Distributions. Keeping the notation of IV. in the last article, let T be a closed surface equipotential with respect either to the joint action of two distributions of matter, M_1 inside T and M_2 outside it, or (when M_2 equals zero) to the action of a single distribution within the surface; and let V_1, V_2, and V be the values of the potential functions due respectively to M_1 alone, to M_2 alone, and to M_1 and M_2 existing together. If a quantity of matter were condensed on T so as to give at every point a surface density equal to $\dfrac{-D_n V}{4\pi}$, the whole quantity of matter on the surface would be

$$\frac{-1}{4\pi}\int D_n V \cdot ds,$$

and this, by [149], is equal in amount to M_1. Let us study the effect of removing M_1 from the inside of T and spreading it in a superficial distribution M_1' over T, so that the surface density at every point shall be $\dfrac{-D_n V}{4\pi}$. In what follows, it is assumed that we have two distributions of matter, one inside the closed surface and the other outside. It is to be carefully noted, however, that by putting M_2 equal to zero in our equations, all our results are applicable to the case where we have an equipotential surface surrounding all the matter, which may be all of one kind or not.

The value, at any point O, of the potential function due to the joint effect of M_2 and the surface distribution M_1', would be

$$V_0 = V_2 - \frac{1}{4\pi}\int \frac{D_n V}{r} \cdot ds.$$

If O is an outside point, we have, by [153],

$$V_0 = V_2 + V_1,$$

so that the effect at any point outside an equipotential surface of a quantity M_1 of matter anyhow distributed inside the surface is the same as that of an equal quantity of matter distributed over the surface in such a way that the superficial density at every point is $\dfrac{-D_n V}{4\pi}$, where V is the value of the potential function due to the joint action of M_1 and any matter (M_2) that may be outside the surface.

If O is an inside point, we have, by [157],

$$V_0 = V_2 + V_, - V_2 = V_,, \qquad [161]$$

which shows that the joint effect of M_2 and M_1' is to give to all points within and upon the surface the same constant value of the potential function which points upon the surface had before M_1 was displaced by M_1'. If, therefore, M_1' and M_2 exist without M_1, there is no force at any point of the cavity shut in by T; or, in other words, the force due to M_1' alone is at all points inside T equal and opposite to that due to M_2.

If M_1 and M_2 exist without M_1', the cavity enclosed by T is, in general, a field of force. M_1' acts as a screen to shield the space within T from the action of M_2.

The surface of M_1' is equipotential with respect to all the active matter, so that there is no tendency of the matter composing the surface distribution to arrange itself in any different manner upon T.

51. The potential function V, due to any distribution of matter whose volume density ρ is everywhere finite, satisfies the following conditions :

(1) V and its first space derivatives are everywhere finite and continuous, and are equal to zero at an infinite distance from the attracting mass.

(2) If R is the distance from the origin of coördinates to the point P,

$$\underset{R = \infty}{\text{limit}} \, (V_P \cdot R) = M,$$

where M is a definite constant.

(3) Except at the surface of the attracting mass, or at some other surface where ρ is discontinuous,

$$\nabla^2 V = - 4\pi\rho,$$

where ρ is to be put equal to zero outside of the attracting mass.

It is easy to show from Green's Theorem that for a given value of ρ as a function of x, y, and z, only one function which will satisfy these three conditions exists.

Suppose, for the sake of argument, that there are two such functions, V and V', and put $u = V - V'$. It is evident that u satisfies conditions (1) and (2), and that $\nabla^2(u) = 0$ except where ρ is discontinuous. Parallel to each surface of discontinuity, and very near to it, draw two surfaces, one on each side, so as to shut in the places where $\nabla^2 u$ is not zero, and draw a spherical surface about the origin, using a radius R large enough to enclose all the surfaces of discontinuity.

If now we apply [151] to that part of the space inside the spherical surface and not shut in by the barriers which we have drawn, and if we notice that each pair of parallel barriers together yields nothing to the surface integral, we shall have

$$\iiint [(D_x u)^2 + (D_y u)^2 + (D_z u)^2] \, dx \, dy \, dz = \int u \cdot D_n u \cdot ds,$$

where the dexter integral is to be extended over the spherical surface only.

If $d\omega$ is the solid angle of the infinitesimal cone which intercepts the element ds from the spherical surface, we have

$$\int u D_n u \, ds = R^2 \int u D_R u \, d\omega.$$

Now since u satisfies condition (2) above, it is easy to show that if we make R grow larger and larger, this surface integral approaches the value zero as a limit, for u approaches the value $\frac{M}{R}$ and $D_R u$ the value $\frac{-M}{R^2}$, so that the whole integral approaches the value $\frac{-4\pi M}{R}$, which, when R is made infinite, approaches the value zero.

If we embrace all space in our sphere, we shall have

$$\iiint (D_x u)^2 + (D_y u)^2 + (D_z u)^2] \, dx \, dy \, dz = 0,$$

whence $\qquad D_x u = 0, \quad D_y u = 0, \quad D_z u = 0.$

Therefore u is constant in all space, and since it is zero at infinity, must be everywhere zero, so that $V = V'$.

52. Thomson's Theorem or Dirichlet's Principle. We will now prove a theorem[*] which is usually called Dirichlet's Principle by Continental writers, but which in English books is generally attributed to Sir W. Thomson. This theorem, in its simplest form, asserts that there always exists one, but no other than this one, function, v, of x, y, z, which (1) is finite, continuous, and single-valued, together with its first space derivatives, throughout a given closed region L; (2) at every point of the region satisfies the equation $\nabla^2 v = 0$; and (3) at every point on the boundary of the region has any arbitrarily assigned value, provided that this can be regarded as the value at that point of a single-valued function which has derivatives finite, continuous, and single-valued all over this boundary.

There is evidently an infinite number of functions which satisfy the first and third conditions. If, for instance, the equation of the bounding surface S of the region is $F(x, y, z) = 0$, and if the value of v at the point (x, y, z) upon this surface is to be $f(x, y, z)$, any function of the form

$$\Phi(x, y, z) \cdot F(x, y, z) + f(x, y, z)$$

would satisfy the third condition, whatever finite function Φ might be.

If we assign to the function to be found a constant value C all over S, $v \equiv C$ will satisfy all three of the conditions given above.

---

[*] Green, *An Essay on the Application of Mathematical Analysis to the Theories of Electricity and Magnetism.* Gauss, *Allgemeine Lehrsätze in Beziehung auf die im verkehrten Verhältnisse des Quadrats der Entfernung wirkenden Anziehungs- und Abstossungskräfte.* Thomson, *Reprint of Papers on Electrostatics and Magnetism.* Dirichlet, *Vorlesungen über die im umgekehrten Verhältniss des Quadrats der Entfernung wirkenden Kräfte.* Also, Thomson and Tait's *Natural Philosophy,* and several papers by Dirichlet in Crelle's Journal and in the Comptes Rendus.

If the sought function is to have different values at different points of S, and if for u in the integral

$$Q = \int \int \int [(D_x u)^2 + (D_y u)^2 + (D_z u)^2] \, dx \, dy \, dz,$$

which is to be extended over the whole of the region, we substitute any one of all the functions which satisfy conditions (1) and (3), the resulting value of Q will be positive. Some one at least of these functions (v) must, however, yield a value of Q which though positive, is so small that no other one can make Q smaller. Let h be an arbitrary constant to which some value has been assigned, and let w be any function which satisfies condition (1) and is equal to zero at all parts of S, then $U = v + hw$ will satisfy conditions (1) and (3), and conversely, there is no function which satisfies these two conditions which cannot be written in the form $U = v + hw$, where h is an arbitrary constant, and w a function which is zero at S and which satisfies condition (1).

Call the minimum value of Q due to v, Q_v, and the value of Q due to U, Q_U, then

$$Q_U = Q_v + 2h \int \int \int (D_x v \cdot D_x w + D_y v \cdot D_y w + D_z v \cdot D_z w) \, dx \, dy \, dz$$

$$+ h^2 \int \int \int [(D_x w)^2 + (D_y w)^2 + (D_z w)^2] \, dx \, dy \, dz,$$

which, since w is zero at the boundary of the region, may be written, by the help of Green's Theorem,

$$Q_U - Q_v = -2h \int \int \int w \nabla^2(v) \, dx \, dy \, dz + h^2 \Omega^2.$$

Now since Q_v is the minimum value of Q, no one of the infinite number of values of $Q_U - Q_v$ formed by changing h and w under the conditions just named can be negative; but if $\nabla^2 u$ were not everywhere equal to zero within L, it would be easy to choose w so that the coefficient of $2h$ in the expression for $Q_U - Q_v$ should not be zero, and then to choose h so that $Q_U - Q_v$ should be negative. Hence $\nabla^2 v$ is equal to zero throughout L, and there always exists at least one function which satisfies the three conditions stated above.

There is only one such function; for if beside v there were another $u = v + hw$, we should have, since the coefficient of h is zero when $\nabla^2(u) = 0$,

$$Q_u = Q_v + h^2 \Omega^2,$$

and, that Q_u may be as small as Q_v, $h\Omega$ must be zero, whence either $h = 0$ or $\Omega = 0$, and if $\Omega = 0$, w is zero. Therefore, $u = v$, and there is only one function which in any given case satisfies all the three conditions given above.

By applying the same reasoning to the space outside a closed surface S and inside a spherical surface of large radius R which is finally made infinite, it is easy to prove that there always exists in the space outside a closed surface S one and only one function v which (1) has a given value at every point of S, (2) satisfies the equation $\nabla^2 v = 0$, (3) together with its first derivatives, is finite and continuous outside S, and (4) is such that the limit, as R becomes infinite, of Rv is a definite, finite constant.

These theorems help us to prove other theorems, of which two are of considerable interest for us.

I. If a function $v = f(x, y, z)$, together with its first space derivatives, is finite and continuous in all space outside a surface S, and outside this surface satisfies the equation $\nabla^2 v = 0$, and if $v\sqrt{x^2 + y^2 + z^2}$ approaches a definite, finite, constant limit as the point (x, y, z) moves away from the origin to infinity, then this function may be considered to be the potential function of a surface distribution of matter upon S.

In order to prove this, we will first apply [160] to v', the function which has on S the same value as v, which inside S is, with its first derivatives, finite and continuous, and which satisfies the equation $\nabla^2 v' = 0$; and use the space inside S as our region. This gives

$$\int v D_n \left(\frac{1}{r}\right) ds = \int \frac{D_n v'}{r} ds,$$

where n refers to the exterior normal of S.

If we now apply [159] to the function v, using as a field the space outside S and within a spherical surface S' of large radius R, drawn about the point O as centre, we shall have

$$4\pi v_0 = -\int v\, D_n\left(\frac{1}{r}\right) ds - \int \frac{D_n v}{r}\, ds + \frac{1}{R^2}\int v\, ds' + \frac{1}{R}\int D_R v \cdot ds',$$

where n is made to refer to the same normal as before by a change of sign in the first two integrals. If now we combine the two equations just obtained, and make R infinite, so that the last two integrals of the second equation shall vanish, we shall have

$$v_0 = +\frac{1}{4\pi}\int (D_n v' - D_n v)\frac{ds}{r},$$

which is the value at an outside point of the potential function due to a superficial distribution of surface density $\frac{1}{4\pi}(D_n v' - D_n v)$ spread upon S.

It is to be noticed that the letter r refers to a point without S in each of the last three equations. Instead of one closed surface we might have several, as it is easy to prove by introducing as many Dirichlet's functions as there are surfaces.

We will state the second theorem, leaving the proof, which is almost identical in form with the one just given, for the reader.

II. If a function $v' = F(x, y, z)$ satisfies the equation $\nabla^2 v' = 0$ throughout the space enclosed by a closed surface S, and within this space, together with its first derivatives, is everywhere finite and continuous, it may be considered to be the potential function within this space of a surface distribution on S.

The superficial density of this distribution will be found to be

$$\frac{1}{4\pi}(D_n v' - D_n v),$$

where v is the function which has the same value on S that v' has, and outside S satisfies the equation $\nabla^2(v) = 0$ and the other conditions given above.

It follows, from these theorems, that we may assign any continuously arranged arbitrary values to the potential function at the different points of a closed surface S, make these values the common values on the surface of the functions v and v', and assert that a distribution of matter on S of surface density

$$\sigma = \frac{1}{4\pi}(D_n v' - D_n v)$$ would give rise to a potential function

having the chosen values on S. In this case v and v' would be the values in regions respectively without and within S of the potential function due to this surface distribution. It is, then, always possible to distribute matter in one and only one way upon a given closed surface so that the value of the potential function due to the matter shall have given values all over the surface.

EXAMPLES.

1. Prove that there always exists one, but no other than this one, function, v, which, together with its first space derivatives, is finite, continuous, and single-valued everywhere within a given region L, has values at the boundary of the region equal to those of an arbitrarily chosen, finite, continuous, and single-valued function, $f(x, y, z)$, and satisfies at every point in L the equation

$$D_x(K \cdot D_x v) + D_y(K \cdot D_y v) + D_z(K \cdot D_z v) = 0,$$

where K is a function positive within L.

2. If the potential function due to a certain distribution of matter is given equal to zero for all space external to a given closed surface S and equal to $\phi(x, y, z)$, where ϕ is a continuous single-valued function zero at all points of S, for all space within S; there is no matter without S, there is a superficial distribution of surface density

$$\sigma = \frac{1}{4\pi}[(D_x\phi)^2 + (D_y\phi)^2 + (D_z\phi)^2]$$

upon S, and the volume density of the matter within S is

$$\rho = -\frac{1}{4\pi}[D_x^2\phi + D_y^2\phi + D_z^2\phi].$$

[Thomson and Tait.]

CHAPTER V.

ELECTROSTATICS.

53. Introductory. Having considered abstractly a few of the characteristic properties of what has been called "the Newtonian potential function," we will devote this chapter to a very brief discussion of some general principles of Electrostatics. By so doing we shall incidentally learn how to apply to the treatment of practical problems many of the theorems that we have proved in the preceding chapters.

In what follows, the reader is supposed to be familiar with such electrostatic phenomena as are described in the first few chapters of treatises on Statical Electricity, and with the hypotheses that are given to explain these phenomena.

Without expressing any opinion with regard to the physical nature of what is called *electrification*, we shall here take for granted that whether it is due to the presence of some substance, or is only the consequence of a mode of motion or of a state of polarization, we may, without error in our results, use some of the language of the old "Two Fluid Theory of Electricity" as the basis of our mathematical work.

The reader is reminded that, among other things, this theory teaches that : —

(1) Every particle of a body which is in its natural state contains, combined together so as to cancel each other's effects at all outside points, equal large quantities of two kinds of *electricity* with properties like those of the positive and negative "matter" described in Section 44.

(2) Electrification consists in destroying in some way the equality between the amounts of the two kinds of electricity which a body, or some part of a body, naturally contains, so that there shall be an excess or *charge* of one kind. If the

charge is of positive electricity, the body is said to be positively electrified ; if the charge is negative, negatively electrified. Either kind of electricity existing uncombined with an equal quantity of the other kind, is called *free* electricity.

(3) When a charged body A is brought into the neighborhood of another body B in its natural state, the two kinds of electricity in every particle of B tend to separate from each other, one being attracted and the other repelled by A's charge, and to move in opposite directions.

In general, a tendency to separation occurs in all parts of the body, whether it is charged or not, where the resultant electric force (the force due to all the free electricity in existence) is not zero. This effect is said to be due to *induction*.

In our work we shall assume all this to be true, and proceed to apply the principles stated in Section 44 to the treatment of problems involving distributions of electricity. We shall find it convenient to distinguish between *conductors*, which offer practically no resistance to the passage of electricity through their substance, and *nonconductors*, which we shall regard as preventing altogether such transfer of electricity from part to part.

54. The Charges on Conductors are Superficial. When electricity is communicated to a conductor, a state of equilibrium is soon established. After this has taken place, there can be no resultant force tending to move any portion of the charge through the substance of the conductor, for, by supposition, the conductor does not prevent the passage of electricity through itself.

Moreover, the resultant electric force must be zero at all points in the substance of a conductor in electric equilibrium ; for if the force were not zero at any point, electricity would be produced by induction at that point, and carried away through the body of the conductor under the action of the inducing force.

From this it follows that the potential function V, due to all the free electricity in existence, must be constant throughout the substance of any single conductor in electric equilibrium, whether or not the conductor be charged, and whether or not there be other charged or uncharged conductors in the neighborhood. Different conductors existing together will in general be at different potentials, but all the points of any one of these conductors will be at the same potential.

Wherever V is constant, $\nabla^2 V = 0$, and hence, by Poisson's Equation, $\rho = 0$, so that there can be no free electricity within the substance of a conductor in equilibrium, and the whole charge must be distributed upon the surface. Experiment shows that we must regard the thickness of charges spread upon conductors as inappreciable, and that it is best to consider that in such cases we have to do with really superficial distributions of electricity, in which the conductor bears a rough analogy to the cavity enclosed by the thin shells of repelling matter described in the preceding chapter.

The surface density at any point of a superficial distribution of electricity shall be taken positive or negative, according as the electricity at that point is positive or negative, and the force which would act upon a unit of positive electricity if it were concentrated at a point P without disturbing existing distributions shall be called "the electric force" or "the strength of the electric field at P."

It is evident, from Sections 45 and 46, that the electric force at a point just outside a charged conductor, at a place where the surface density of the charge is σ, is $4\pi\sigma$, and that this is directed outwards or inwards, according as σ is positive or negative.

In other words, $D_n V$, the derivative of the potential function in the direction of the exterior normal, is equal to $-4\pi\sigma$, and the value of V at a point P just outside the conductor is greater or less than its value within the conductor, according as the surface density of the conductor's charge in the neighborhood of P is negative or positive.

It is to be carefully noted that, although the surface of a conductor must always be equipotential, the superficial density of the conductor's charge need not be the same at all parts of the surface. We shall soon meet with cases where the electricity on a conductor's surface is at some points positive and at others negative, and with other cases where the sign of the potential function inside and on a conductor is of opposite sign to the charge.

It is evident, from the work of Section 47, that the resistance per unit of area which the nonconducting medium about a conductor has to exert upon the conductor's charge to prevent it from flying off, is, at a part where the density is σ, $2\pi\sigma^2$.

55. General Principles which follow directly from the Theory of the Newtonian Potential Function. If two different distributions of electricity, which have the same system of equipotential surfaces throughout a certain region, be superposed so as to exist together, the new distribution will have the same equipotential surfaces in that region as each of the components. For, if V_1 and V_2, the potential functions due to the two components respectively, be both constant over any surface, their sum will be constant over the same surface.

Two distributions of electricity, which have densities everywhere equal in magnitude but opposite in sign, have the same system of equipotential surfaces, and, if superposed, have no effect at any point in space.

Two distributions of electricity, arranged successively on the same conductor so that at every point the density of the one is m times that of the other, have the same system of equipotential surfaces, and the potential function due to the first is everywhere m times as great as that due to the second.

If the whole charge of a conductor which is not exposed to the action of any electricity except its own is zero, the superficial density must be zero at all points of the surface, and the conductor is in its natural state. For if σ is not everywhere zero, it must be in some places positive and in others negative; and, according to the work of the last section, the potential function V, due to this charge, must have, somewhere outside the conductor, values higher and lower than V_0, its value in the conductor itself. But this would necessitate somewhere in empty space a value of the potential function not lying between V_0 and 0, the value at infinity; that is, a maximum in empty space if V_0 is positive, and a minimum if V_0 is negative; which is absurd.

A system of conductors, on each of which the charge is null, must be in the natural state if exposed to the action of no outside electricity. For, by applying the reasoning just used to that conductor in which the potential function is supposed to have the value most widely different from zero, we may show that the surface density all over the conductor is zero, so that no influence is exercised on outside bodies; and then, supposing this conductor removed, we may proceed in the same way with the system made up of the remaining conductors.

If a charge M of electricity, when given to a conductor, arranges itself in equilibrium so as to give the surface density $\sigma = f(x, y, z)$ and to make the potential function $V_0 = \int \frac{\sigma \, ds}{r}$ constant within the conductor, a charge $-M$, if arranged on the conductor so as to give at every point the density $-\sigma = -f(x, y, z)$ would be in equilibrium, for it would give everywhere the potential function $\int \frac{-\sigma \cdot ds}{r} = -V_0$, and this is constant wherever V_0 is constant.

Only one distribution of the same quantity of electricity M on the same conductor, removed from the influence of all other electricity, is possible; for, suppose two different values of surface density possible, $\sigma_1 = f_1(x, y, z)$ and $\sigma_2 = f_2(x, y, z)$, then $-\sigma_2 = -f_2(x, y, z)$ is a possible distribution of the charge $-M$. Superpose the distribution $-\sigma_2$ upon the distribution σ_1 so that the total charge shall be equal to zero: then the surface density at every point is $\sigma_1 - \sigma_2$, and this must be zero by what we have just proved, so that $\sigma_1 = \sigma_2$.

Since we may superpose on the same conductor a number of distributions, each one of which is by itself in equilibrium, it is easy to see that if the whole quantity of electricity on any conductor be changed in a given ratio, the density at each point will be changed in the same ratio.

56. Tubes of Force and their Properties. We have seen that a unit of positive electricity concentrated at a point P just outside a conductor would be urged away from the conductor or drawn towards it, according as that point on the conductor which is nearest P is positively or negatively electrified. If we regard lines of force drawn in an electric field as generated by points moving *from* places of higher potential *to* places of lower potential, we may say that a line of force *proceeds from* every point of a conductor where the surface density is positive, and that a line of force *ends at* every point of a conductor where the surface density is negative. No line of force either leaves or enters a conductor at a point where the surface density is zero, and no line of force can start at one point of a conductor where the electrification is positive and return to the same conductor at a point where the electrification is negative. No line of force can proceed from one conductor at a point electrified in any way and enter another conductor at a point where the electrification has the same name as at the starting-point. A line of force never cuts through a conductor so as to come out at the other side, for the force at every point inside a conductor is zero.

Lines and tubes of force are sometimes called in electrostatics lines and tubes of " induction."

When a tube of force joins two conductors, the charges Q_1, Q_2 of the portions S_1, S_2 which it cuts from the two surfaces are

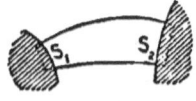

Fig. 38.

made up of equal quantities of opposite kinds of electricity. For if we suppose the tube of force to be arbitrarily prolonged

and closed at the ends inside the two conductors, the surface
integral of normal force taken over the box thus formed is zero,
for the part outside the conductors yields nothing, since the resultant force is tangential to it, and there is no resultant force
at any point inside a conductor.   It follows, from Gauss's
Theorem, that the whole quantity of electricity $(Q_1 + Q_2)$ inside
the box must be zero, or $Q_1 = -Q_2$, which proves the theorem.
If σ_1 and σ_2 are the average values of the surface densities of
the charges on S_1 and S_2 respectively, we have $\sigma_1 S_1 = Q_1$ and
$\sigma_2 S_2 = Q_2$, whence

$$\sigma_2 = -\sigma_1 \frac{S_1}{S_2} \qquad [162]$$

The integral taken over any surface, closed or not, of the
force normal to that surface is called by some writers the *flow
of force* across the surface in question, and by others the *induction* through this surface.

If we apply Gauss's Theorem to a box shut in by a tube
of force and the portions S_1, S_2 which it cuts from any two
equipotential surfaces, we shall have, if the box contains no
electricity,

$$F_2 S_2 - F_1 S_1 = 0, \qquad [163]$$

where F_1 and F_2 are the average values, over S_1 and S_2 respectively, of the normal force taken in the same direction (that in
which V decreases) in both cases.   In other words, the flow of
force across all equipotential sections of a tube of force containing no electricity is the same, or the average force over an
equipotential section of an empty tube of force is inversely proportional to the area of the section.

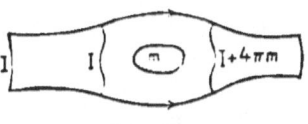

Fig. 39.

When a tube of force encounters a quantity m of electricity
(Fig. 39), the flow of force through the tube on passing this

electricity is increased by $4\pi m$. If, however, the tube encounters a conductor large enough to close its end completely, a charge m will be found on the conductor just sufficient to reduce to zero the flow of force (I) through the tube. That is,

$$m = -\frac{I}{4\pi}.$$

It is sometimes convenient to consider an electric field to be divided up by a system of tubes of force, so chosen that the flow of force across any equipotential surface of each tube shall be equal to 4π. Such tubes are called *unit tubes;* for wherever one of them abuts on a conductor, there is always the unit quantity of electricity on that portion of the conductor's surface which the tube intercepts. In some treatises on electricity the term "line of force" is used to represent a unit tube of force, as when a conductor is said to cut a certain number of "lines of force."

It is evident that m unit tubes abut on a surface just outside a conductor charged with m units of either kind of electricity, if the superficial density of the charge has everywhere the same sign. These tubes must be regarded as *beginning* at the conductor if m is positive, and as *ending* there if m is negative. If a conductor is charged at some places with positive electricity and at others with negative electricity, tubes of force will begin where the electrification is positive, and *others* will end where the electrification is negative.

It is evident that no tube of force can return into itself.

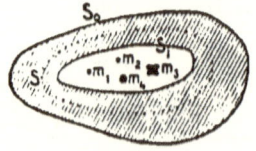

Fig. 40.

57. Hollow Conductors. When the nonconducting cavity, shut in by a hollow conductor K (Fig. 40), contains quantities of electricity (m_1, m_2, m_3, etc., or $\sum m$) distributed in any way, but insulated from K, there is induced on the walls of the cavity a charge of electricity algebraically equal in quantity, but opposite in sign, to the algebraic sum of the electricity within the cavity.

Call the outside surface of the conductor S_o and its charge M_o, the boundary of the cavity S_i and its charge M_i, and surround the cavity by a closed surface S, every point of which lies within the substance of the conductor, where the resultant force is zero. Now the surface integral of normal force taken over S is zero, so that, according to Gauss's Theorem, the algebraic sum of the quantities of electricity within the cavity and upon S_i is zero. That is,

$$M_i + m_1 + m_2 + m_3 + \cdots = M_i + \sum (m) = 0, \qquad [164]$$

and this is our theorem, which is true whatever the charge on S_o is, and whatever distribution of free electricity there may be outside K. If the distribution of the electricity within the cavity be changed by moving m_1, m_2, etc., to different positions, the *distribution* of M_i on S_i will in general be changed, although its value remains unchanged.

If K has received no electricity from without, its total charge must be zero; that is,

$$M_o = - M_i = \sum (m).$$

If a charge algebraically equal to M be given to K,

$$M_o = M - M_i.$$

The combined effect of $\sum (m)$, the electricity within the cavity, and M_i, the electricity on the walls of the cavity, is at all points without S_i absolutely null. For, if we apply [153] to S, any surface drawn in the conductor so as to enclose S_i, we shall have $D_n V$ everywhere zero, since the potential function is constant within the conductor; this shows that V_1, the potential function due to all the electricity within S, must be zero at all points without S; but S may be drawn as nearly coincident with S_i as we please. Hence our theorem, which shows that, so far as the value of the potential function in the substance of the conductor or outside it, and so far as the arrangement of M_o and of M', any free electricity there may be outside K, are concerned, M_i and $\sum (m)$ might be removed together without changing anything. The potential function at all points outside S_i is to be found by considering only M and M'.

If S_i happens to be one of the equipotential surfaces of $\sum (m)$ considered by itself, M_i will be arranged in the same way as a charge of the same magnitude would arrange itself on a conductor whose outside surface was of the shape S_i, if removed from the action of all other free electricity.

The potential function (V_2) due to M_o and M' is constant everywhere within S_o; for if we apply [157] to a surface S, drawn within the substance of the conductor as near S_o as we like, we shall have

$$V_s - V_0 = 0,$$

which proves the theorem.

The potential function within the cavity is equal to $V_2 + V_1$, where V_1 is the potential function due to M_i and $\sum (m)$. Of these, V_2 is, as we have seen, constant throughout K and the cavity (Section 31) which it encloses, while V_1 has different values in different parts of the cavity, and is zero within the substance of the conductor.

Suppose now that, by means of an electrical machine, some of the two kinds of electricity existing combined together in a conductor within the cavity be separated, and equal quantities (q) of each kind be set free and distributed in any manner within the cavity.

The value of V_1 within the cavity will probably be different from what it was before, but V_2 will be unchanged; for the quantity of matter in the cavity is unchanged, being now, algebraically considered,

$$\sum (m) + q - q = \sum (m),$$

so that M_i is unchanged, although it may have been differently arranged on S_i, in order to keep the value of V_1 zero within the substance of the conductor. If now a part of the free electricity in the cavity be conveyed to S_i in some way, the substance of the conductor will still remain at the same potential as before. For, if l units of positive electricity and n units of negative electricity be thus transferred to S_i, the whole quantity of free electricity within the cavity will be $\sum (m) - l + n$, and that on S_i will be $M_i + l - n$: but these are numerically equal, but opposite in sign, and the charge on S_i, if properly arranged, suffices, without drawing on M_0 to reduce to zero the value of V_1 in K. Since M_0 and M' remain as before, V_2 is unchanged, and the conductor is at the same potential as before. So long as no electricity is introduced into the cavity from *without* K, no electrical charges within the cavity can have any effect outside S_i.

Most experiments in electricity are carried on in rooms, which we can regard as hollows in a large conductor, the earth. V_2, the value of the potential function in the earth and the walls of the room, is not changed by anything that goes on inside the room, where the potential function is $V = V_1 + V_2$. Since we are generally concerned, not with the absolute value of the potential function, but only with its variations within the room, and since V_2 remains always constant, it is often convenient to disregard V_2 altogether, and to call V_1 the value of the potential function inside the room. When we do this we must remember that we are taking the value of the potential function in the earth as an arbitrary zero, and that the value of V_1 at a point in the room really measures only the difference between the values of the potential function in the earth and at the point in question. When a conductor A in the room is connected with the walls of the room by a wire, the value of V_1 in A is, of course, zero, and A is said to have been *put to earth*.

58. Induced Charge on a Conductor which is put to Earth. Suppose that there are in a room a number of conductors, viz. : A_1 charged with M_1 units of electricity, and A_2, A_3, A_4, etc., connected with the walls of the room, and therefore at the potential of the earth, which we will take for our zero. If the potential function has the value p_1 inside A_1, every point in the room outside the conductors must have a value of the potential function lying between p_1 and 0, else the potential function must have a maximum or a minimum in empty space. If p_1 is positive, there can be no positive electricity on the other conductors ; for if there were, lines of force must start from these conductors and go to places of lower potential ; but there are no such places, since these conductors are at potential zero, and all other points of the room at positive potentials. In a similar way we may prove that if p_1 is negative, the electricity induced on the other conductors is wholly positive.

Now let us apply [158] to a spherical surface, drawn so as to include A_1 and at least one of the other conductors, but with radius a so small that some parts of the surface shall lie within the room. If we take the point O at the centre of this surface, we shall have

$$4\pi V_2 = \frac{1}{a}\int D_r V \cdot ds + \frac{1}{a^2}\int V ds. \qquad [165]$$

If M is the whole quantity of electricity within the spherical surface, there must be a quantity $-M$ outside the surface, either on the walls of the room or on conductors within the room. The value at O of the potential function, V_2, due to the electricity without the sphere, is less in absolute value than $-\dfrac{M}{a}$, for it could only be as great as this if all the electricity outside the sphere were brought up to its surface.

By Gauss's Theorem,

$$\int D_r V \cdot ds = -4\pi M,$$

therefore, $\int V ds = 4\pi a [M + a V_2].$ [166]

Now, if M_1 is positive, the integral is positive, for all parts of the spherical surface within the room yield positive differentials, and all other parts zero, so that the second side of the equation is positive. But $a V_2$ is of opposite sign to M, and is less in absolute value; hence, M is positive, and the total amount of negative electricity induced on the other conductors within the spherical surface by the charge on A_1, is numerically less than this charge, unless some one of these conductors surrounds A_2; in which case the induced charge comes wholly on this conductor, while the other conductors, and the walls of the room, are free. Some of the tubes of force which begin at A_1 end on the walls of the room, provided these latter can be reached from A_1 without passing through the substance of any conductor.

59. Coefficients of Induction and Capacity. If a number of insulated conductors, A_2, A_3, A_4, etc., are in a room in the presence of a conductor A_1 charged with M_1 units of electricity, the whole charge on each is zero; but equal amounts of positive and negative electricity are so arranged by induction on each, that the potential function is constant throughout the substance of every one of the conductors.

Let the values of the potential functions in the system of conductors be p_1, p_2, p_3, p_4, etc. Since each conductor except A_1 is electrified, if at all, in some places with positive electricity, and in others with negative electricity, some lines of force must start from, and others end at, every such electrified conductor, so that there must be points in the air about each conductor at lower and at higher potentials than the conductor itself. But the value of the potential function in the walls of the room is zero, and there can be no points of maximum or minimum potential in empty space; so that p_1 must be that value of the potential function in the room most widely different from zero, and p_2, p_3, p_4, etc., must have the same sign as p_1.

The reader may show, if he likes, that both the negative part and the positive part of the zero charge of any conductor, except A_1, is less than M_1.

Let p_{11} be the value of the potential function in a conductor A_1 charged with a single unit of electricity, and standing in the presence of a number of other conductors all uncharged and insulated. Then if p_{12}, p_{13}, p_{14}, etc., are, under these circumstances, the values of the potential functions in the other conductors, A_2, A_3, A_4, etc., the potential functions in these conductors will be $M_1 p_{12}$, $M_1 p_{13}$, $M_1 p_{14}$, etc., if A_1 be charged with M_1 units of electricity instead of with one unit. This is evident, for we may superpose a number of distributions which are singly in equilibrium upon a set of conductors, and get a new distribution in equilibrium where the density is the sum of the densities of the component distributions, and the value of the resulting potential function the sum of the values of their potential functions.

If A_1 be discharged and insulated, and a charge M_2 be given to A_2, the values of the potential functions in the different conductors may be written

$$M_2 p_{21}, \quad M_2 p_{22}, \quad M_2 p_{23}, \quad M_2 p_{24}, \quad \text{etc.}$$

If now we give to A_1 and A_2 at the same time the charges M_1 and M_2 respectively, and keep the other conductors insulated, the result will be equivalent to superposing the second distribution, which we have just considered, upon the first, and the conductors will be respectively at potentials,

$$M_1 p_{11} + M_2 p_{21}, \quad M_1 p_{12} + M_2 p_{22}, \quad M_1 p_{13} + M_2 p_{23}, \quad \text{etc.} \quad [167]$$

If all the conductors are simultaneously charged with quantities M_1, M_2, M_3, M_4, etc., of electricity respectively, the value of the potential function on A_k will be

$$V_k = M_1 p_{1k} + M_2 p_{2k} + M_3 p_{3k} + \cdots + M_k p_{kk} + M_n p_{nk}, \quad [168]$$

Writing this in the form $V_k = a_k + M_k p_{kk}$, we see that if the charges on all the conductors except A_k be unchanged, a_k will be constant, and that every addition of $\dfrac{1}{p_{kk}}$ units of electricity to the charge of A_k raises the value of the potential function in it by unity. If we solve the n equations like [168] for the charges, we shall get n equations of the form

$$M_k = V_1 q_{1k} + V_2 q_{2k} + V_3 q_{3k} + \cdots + V_k q_{kk} + \cdots + V_n q_{nk}, \quad [169]$$

where the q's are functions of the p's.

If all the conductors except A_k are connected with the earth, $M_k = V_k q_{kk}$, and q_{kk} is evidently the charge which, under these circumstances, must be given to A_k in order to raise the value of the potential function in it by unity. It is to be noticed that q_{kk} and $\dfrac{1}{p_{kk}}$ are in general different.

The charge which must be given to a conductor when all the conductors which surround it are in communication with the earth, in order to raise the value of the potential function within that conductor from zero to unity, shall be called the *capacity* of the conductor. It is evident that the capacity of a conductor thus defined depends upon its shape and upon the shape and position of the conductors in its neighborhood.

60. Distribution of Electricity on a Spherical Conductor. Considerations of symmetry show that if a charge M be given to a conducting sphere of radius r, removed from the influence of all electricity except its own, the charge will arrange itself uniformly over the surface, so that the superficial density shall be everywhere $\sigma = \dfrac{M}{4 \pi r^2}$.

The value, at the centre of the sphere, of the potential function due to the charge M on the surface is $\dfrac{M}{r}$, and, since the potential function is constant inside a charged conductor, this must be the value of the potential function throughout the sphere. If M is equal to r, $\dfrac{M}{r} = 1$: hence the capacity of a spherical conductor removed from the influence of all electricity except its own, is numerically equal to the radius of its surface.

61. Distribution of a Given Charge on an Ellipsoid. It is evident from the discussion of homœoids in Chapter I. that a charge of electricity arranged (on a conductor) in the form of a shell, bounded by ellipsoidal surfaces similar to each other (and to the surface of the conductor), and similarly placed, would be in equilibrium if the conductor were removed from the action of all electricity except its own. We may use this principle to help us to find the distribution of a given charge on a conducting ellipsoid.

Let us consider a shell of homogeneous matter bounded by two similar, similarly placed, and concentric ellipsoidal surfaces, whose semi-axes shall be respectively a, b, c, and $(1+a)a$, $(1+a)b$, $(1+a)c$. If any line be drawn from the centre of the shell so as to cut both surfaces, the tangent planes to these two surfaces at the points of intersection will be parallel, and the distance between the planes is pa, where p is the length of the perpendicular let fall from the centre upon the nearer of the planes.

If ρ is the volume density of the matter of which the shell is composed, the mass of the shell is $M = \frac{1}{3}\pi abc\left[(1+a)^3 - 1\right]\rho$, and the rate at which the matter is spread upon the unit of surface is, at any point, $\sigma = \rho\delta$, where δ is the thickness of the shell measured on the line of force which passes through the point in question. Eliminating ρ from these equations, we have

$$\sigma = \frac{M\delta}{4\pi abc\left[a + a^2 + \frac{1}{3}a^3\right]}. \qquad [170]$$

If, now, in accordance with the hypothesis that the thickness of the electric charge on a conductor is inappreciable, we make a smaller and smaller, noticing that δ differs from pa by an infinitesimal of an order higher than the first, we shall have for a strictly surface distribution,

$$\sigma = \frac{Mp}{4\pi abc}. \qquad [171]$$

If the equation of the surface of the ellipsoidal conductor is

$$\frac{x^2}{a^2} + \frac{y^2}{b^2} + \frac{z^2}{c^2} = 1,$$

we have

$$\frac{1}{p} = \sqrt{\frac{x^2}{a^4} + \frac{y^2}{b^4} + \frac{z^2}{c^4}},$$

and

$$\frac{c}{p} = \sqrt{c^2\left(\frac{x^2}{a^4} + \frac{y^2}{b^4}\right) + 1 - \frac{x^2}{a^2} - \frac{y^2}{b^2}}.$$

This last expression shows that, as c is made smaller and smaller, σ approaches more and more nearly the value

$$\frac{M}{4\pi ab\sqrt{1 - \frac{x^2}{a^2} - \frac{y^2}{b^2}}}, \qquad [172]$$

and this gives some idea of the distribution on a thin elliptical plate whose semi-axes are a and b.

For a circular plate, we may put $a = b$ in the last expression, which gives

$$\frac{M}{4\pi a\sqrt{a^2 - r^2}} \qquad [173]$$

for the surface density at a point r units distant from the centre of the plate.

The charge M distributed according to this law on both sides of a circular plate of radius a raises the plate to potential

$$V = \frac{M}{a}\int_0^a \frac{dr}{\sqrt{a^2 - r^2}} = \frac{\pi M}{2a},$$

so that the capacity of the plate is

$$\frac{2a}{\pi}. \qquad [174]$$

62. Spherical Condensers. If a conducting sphere A of radius r (Fig. 41) be surrounded by a concentric spherical conducting shell B of radii r, and r_0 and charged with m units of electricity while B is uncharged and insulated, we shall have

(1) the charge m uniformly distributed upon S, the surface of the sphere ;

(2) an induced charge $-m$ (Section 57) uniformly distributed upon S_i, the inner surface of B ;

(3) a charge $+m$ (since the total charge of B is zero) uniformly distributed on S_o, the outer surface of B.

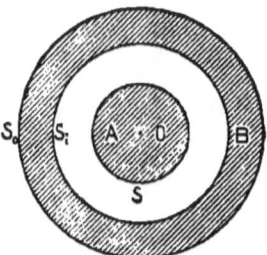

FIG. 41.

The value at the centre of the sphere of the potential function due to all these distributions is $V_A = \dfrac{m}{r} - \dfrac{m}{r_i} + \dfrac{m}{r_o}$, and this is the value of V throughout the conducting sphere. The value of the potential function in B is $V_B = \dfrac{m}{r_o}$.

If now a charge M be communicated to B, this will add itself to the charge m already existing on S_o, and the charge on S_i will be undisturbed. The values of the potential functions in the conductors are now

$$ V_A = \frac{m}{r} - \frac{m}{r_i} + \frac{m+M}{r_o}, \ \text{ and } \ V_B = \frac{m+M}{r_o}. $$

If now B be connected with the earth so as to make $V_B = 0$, the charges on S and S_i will be undisturbed, but the charge on S_o will disappear. V_A is now equal to $\dfrac{m}{r} - \dfrac{m}{r_i}$.

If A were uncharged, and B had the charge M, this charge would be uniformly distributed upon S_o, for, since the whole charge on S is zero, the whole charge on S_i must be zero also. It is easy to see that S and S_i must both be in a state of nature, for if not, lines of force must start from S and end at S_i, and others start at S_i and end at S, which is absurd.

If A were put to earth by means of a fine insulated wire passing through a tiny hole in B, and if B were insulated and charged with M units of electricity, we should have a charge x on S, a charge $-x$ on S_i, and a charge $M+x$ on S_o. To find x, we need only remember that $V_A = \dfrac{x}{r} - \dfrac{x}{r_i} + \dfrac{x}{r_o} + \dfrac{M}{r_o} = 0$, whence x may be obtained.

If B be put to earth, and A be connected by means of the fine wire just mentioned, with an electrical machine which keeps its prime conductor constantly at potential V_1, A will receive a charge y and will be put at potential V_1. To find y, it is to be noticed that there is a charge $-y$ on S_i, and no charge on S_o, which is put to earth. $V_A = \dfrac{y}{r} - \dfrac{y}{r_i} = V_1$, whence y may be obtained.

If $r = 99$ millimeters and $r_i = 100$ millimeters, $y = 9900 \ V_1$.

If a sphere, equal in size to A but having no shell about it, were connected with the same prime conductor, it too would receive a charge z sufficient to raise it to potential V_1, and z would be determined by the equation $V_1 = \dfrac{z}{r}$. If $r = 99$, we have $z = 99 \ V_1$; hence we see that A, when surrounded by B at potential zero, is able to take one hundred times as great a charge from a given machine as it could take if B were removed. In other words, B increases A's capacity one hundred fold. A and B together constitute what is called a *condenser*.

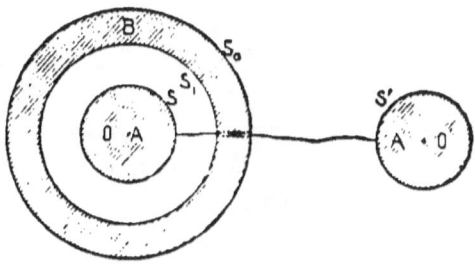

Fig. 42.

If A of the condenser AB, both parts of which are supposed uncharged, be connected by a fine wire (Fig. 42) with a sphere

A' which has the same radius as A, and is charged to potential V_1, A and A' will now be at the same potential $[V_2]$, and A will have the charge x, and A' the charge y. The total quantity of electricity on A' at first was rV_1, so that $x + y = rV_1$, and

$$V_2 = \frac{y}{r} = \frac{x}{r} - \frac{x}{r_i} + \frac{x}{r_o},$$

whence x and y may be found.

The reader may study for himself the electrical condition of the different parts of two equal spherical condensers (Fig. 43),

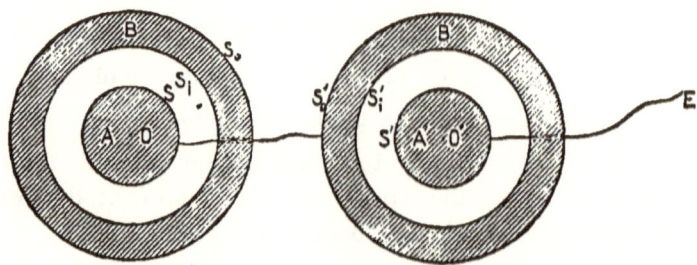

FIG. 43.

of which the outer surface S_o of one is connected with an electric machine at potential V_1, and the inside of the other, S', is connected with the earth. The two condensers, which are supposed to be so far apart as to be removed from each other's influence, illustrate the case of two Leyden jars arranged in cascade.

63. Condensers made of Two Parallel Conducting Plates. Suppose two infinite conducting planes A and B to be parallel to each other at a distance a apart; choose a point of the plane A for origin, and take the axis of x perpendicular to the planes, so that their equations shall be $x = 0$ and $x = a$. Let the planes be charged and kept at potentials V_A and V_B respectively. It is evident from considerations of symmetry that the potential function at the point P between the two planes depends only upon P's x coördinate, so that

$$D_y V = 0, \quad D_z V = 0, \quad D_y^2 V = 0, \quad D_z^2 V = 0.$$

Laplace's Equation gives, then,

$$D_x^2 V = 0,$$

whence $D_x V = C$, and $V = Cx + D$.

If $x = 0$, $V = V_A$; and if $x = a$, $V = V_B$; so that

$$V = (V_B - V_A)\frac{x}{a} + V_A, \text{ and } D_x V = \frac{V_B - V_A}{a}.$$

The lines of force are parallel between the planes, and the surface densities of the charges on A and B are

$$\frac{V_A - V_B}{4\pi a} \text{ and } \frac{V_B - V_A}{4\pi a} \text{ respectively.}$$

If we take a portion of area S out of the middle of each plate, there will be a quantity of electricity on S_A equal to $\dfrac{S(V_A - V_B)}{4\pi a}$, and an equal quantity of the other kind of electricity on S_B. The force of attraction between S_A and S_B will be $2\pi\sigma^2 \cdot S$, or

$$\frac{S}{8\pi}\frac{(V_B - V_A)^2}{a^2}.$$

If S_B be put to earth, the charge that must be given to S_A in order to raise it to potential unity is

$$\frac{S}{4\pi a}.$$

In other words, the capacity of S_A is inversely proportional to the distance between the plates.

In the case of two thin conducting plates placed parallel to and opposite each other, at a distance small compared with their areas, the lines of force are practically parallel except in the immediate vicinity of the edges of the plates;* and we may infer

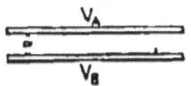

Fig. 44.

* See Maxwell's *Treatise on Electricity and Magnetism*, Vol. I. Fig. XII.

from the results of this section that the capacity of a condenser consisting of two parallel conducting plates of area S, separated by a layer of air of thickness a, when one of its plates is put to earth is very approximately $\dfrac{S}{4\pi a}$ for large values of $\dfrac{S}{a}$.

64. Capacity of a Long Cylinder surrounded by a Concentric Cylindrical Shell. In the case of an infinite, conducting cylinder of radius r_i, kept at potential V_i and surrounded by a concentric conducting cylindrical shell of radii r_o and r', kept at potential V_o, we have symmetry about the axis of the cylinder, so that $D_\phi V = 0$, and Laplace's Equation reduces to the form

$$D_r^2 V + \frac{D_r V}{r} = 0,$$

whence, for all points of empty space between the cylinder and its shell,
$$V = C + D \log r.$$

But $V = V_i$ when $r = r_i$, and $V = V_o$ when $r = r_o$,

hence
$$V = \frac{V_i \log \dfrac{r_o}{r} + V_o \log \dfrac{r}{r_i}}{\log \dfrac{r_o}{r_i}}, \qquad [183]$$

and
$$D_r V = \frac{(V_o - V_i)}{\log \dfrac{r_o}{r_i}} \cdot \frac{1}{r}.$$

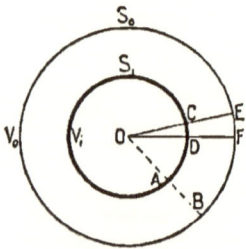

Fig. 45.

The surface densities of the electricity on the outer surface of the cylinder and the inner surface of the shell are respectively

$$\frac{V_i - V_o}{4\pi r_i \log \frac{r_o}{r_i}} \quad \text{and} \quad \frac{V_o - V_i}{4\pi r_o \log \frac{r_o}{r_i}},$$

so that the charge on the unit of length of the cylinder is $\dfrac{V_i - V_o}{2 \log \frac{r_o}{r_i}}$, and the charge on the corresponding portion of the inner surface of the shell is the negative of this. We may find the capacity of the unit length of the cylinder by putting $V_o = 0$ and $V_i = 1$, whence capacity $= \dfrac{1}{2 \log \frac{r_o}{r_i}}$.

If r_o in this expression is made very large, the capacity of the cylinder will be very small.

In the case of a fine wire connecting two conductors, r_i will be very small, and there will be no conducting shell nearer than the walls of the room, so that the capacity of such a wire is plainly negligible.

65. Specific Inductive Capacity. In all our work up to this time we have supposed conductors to be separated from each other by electrically indifferent media, which simply prevent the passage of electricity from one conductor to another. We have no reason to believe, however, that such media exist in nature. Experiment shows, for instance, that the capacity of a given spherical condenser depends essentially upon the kind of insulating material used to separate the sphere from its shell, so that this material, without conducting electricity, modifies the action of the charges on the conductors. Insulators, when considered as transmitting electric action, are sometimes called *dielectrics*.

Whatever may really be the physical natures of the substances, such as glass, paraffine, ebonite, etc., which we commonly use as insulators, it has been shown that their behavior would be fairly well accounted for on the supposition that they arc made up of truly insulating matter in which arc imbedded, at little distances from one another, small, conducting particles.   It is evident that every such particle, if lying in a field of force, would be polarized ; that is, one part would be charged positively by induction, and the part most remote from this would be charged negatively, and that these induced charges would have some influence in determining the values of the potential function at points in the dielectric and in the conductors adjacent to it.

Using the notation of Section 62, let the part A of a spherical condenser be charged with m units of positive electricity and separated from the part B, which is put to earth, by a spherical shell of radii r and r_i made up of a given dielectric. Let us first ask ourselves what the effect of the dielectric would be if it consisted of extremely thin concentric conducting spherical shells separated by extremely thin insulating spaces.   It is evident that in this case we should have a quantity $-m$ of electricity induced on the inside of the innermost shell, a quantity $+m$ on the outside of this shell, a quantity $-m$ on the inner surface of the next shell, a quantity $+m$ on the outside of this shell, and so on.   If there were n such shells in the dielectric layer, and $n + 1$ spaces, and if δ were the distance from the inner surface of one shell to the inner surface of the next, and $\lambda\delta$ the thickness of each shell, the value, at the centre of A, of the potential function due to the charges on these shells, would be

$$V_A' = m\left[\frac{1}{r+\delta} - \frac{1}{r-\lambda\delta+\delta} + \frac{1}{r+2\delta} - \frac{1}{r-\lambda\delta+2\delta}\right.$$
$$\left. + \cdots + \frac{1}{r+n\delta} - \frac{1}{r-\lambda\delta+n\delta}\right]$$
$$= -m\lambda\delta\left[\frac{1}{(r+\delta)(r-\lambda\delta+\delta)} + \frac{1}{(r+2\delta)(r-\lambda\delta+2\delta)} + \cdots\right].$$

This quantity lies between

$$G = -m\lambda\delta\sum_{k=1}^{k=n}\frac{1}{(r+k\delta)^2} \text{ and } H = -m\lambda\delta\sum_{k=0}^{k=n}\frac{1}{(r+k\delta)^2}:$$

but these differ from each other by less than $\epsilon = m\lambda\delta\,\dfrac{r_i^2 - r^2}{r^2 r_i^2}$, so that $-m\lambda\displaystyle\int_r^{\cdot r_i - (1-\lambda)\delta}\dfrac{dx}{x^2}$, which is easily seen to lie between G and H, differs from V_A' by less than ϵ. If, then, δ is very small in comparison with r and r_i, V_A' differs from $m\lambda\left(\dfrac{1}{r_i} - \dfrac{1}{r}\right)$ by an exceedingly small fraction of its own value.

This shows that the effect, at the centre of A, of such a system of conducting shells as we have imagined would be practically the same as if a charge $-m\lambda$ were given to the inner surface of the dielectric, and a charge $+m\lambda$ to its outer surface, while the charges on the surfaces of the thin shells within the mass of the dielectric were taken away. That is, the value of the potential function in A would be

$$m(1 - \lambda)\left(\frac{1}{r} - \frac{1}{r'}\right) \text{ instead of } m\left(\frac{1}{r} - \frac{1}{r'}\right).$$

Such a system of shells introduced into what we have hitherto supposed to be the electrically inert insulating matter between the two parts of a spherical condenser would increase the capacity of the condenser in the ratio of 1 to $1 - \lambda$. It is to be noticed that λ is a proper fraction: $\lambda = 0$ and $\lambda = 1$ would correspond respectively to a perfect insulator and to a perfect conductor.

As Dr. E. H. Hall has suggested to me, the result given above might be easily obtained by computing* the amount of work done in moving a unit particle of electricity (supposed to be concentrated at a point, and not to disturb existing distributions) from A to B. It is easy to see that the force at any point in the mass of one of the thin conducting shells would be zero, and that the force at any point in the space between two shells would be exactly the same as if there were no shells in the dielectric. We have no reason to think that there are any such differences between the values of the force at contiguous points in the dielectric as this would indicate, and the conception of the thin shells has been introduced only

---

* Mascart et Joubert, *Leçons sur l'Électricité*, § 124.

because the effect of these shells can be more easily computed than that of a number of discrete particles.

When, however, the dielectric between the parts of a spherical condenser is supposed to contain not a system of continuous shells, but a number of separate conducting particles, these are often regarded as forming a series of concentric layers, and it is assumed that the sum of the charges induced on the inner sides of the particles in the innermost layer is $-\lambda'm$, where λ' is a proper fraction, larger or smaller in different dielectrics according as the particles are nearer together or farther apart, and that the inner surfaces of all the other layers have each the same charge, and the outer surface of every layer the corresponding positive charge $+\lambda'm$. The effect of this kind of dielectric, if made to replace a perfect insulator in our calculations, would be to increase the capacity of the condenser in the ratio 1 to $1-\mu$, where $\mu = \lambda'\lambda$, and it is evident that the same effect might be produced by a charge $-\mu m$ on that surface of the dielectric which touches A, and a charge $+\mu m$ on that surface which is in contact with B.

Experiment shows that dielectrics used to separate and to surround charged conductors behave, in many respects, as if every surface in contact with a conductor had a charge opposite in sign to that of the conductor, and in absolute value μ times as great, μ being less than unity, and constant for any one dielectric. That is, if the dielectric separating from each other a number of conductors be displaced by another, the capacities of all the conductors will be changed in the same ratio, depending only upon the natures of the two dielectrics.

The ratio of the fraction $\dfrac{1}{1-\mu}$, in the case of any dielectric to the same fraction in the case of air, for which μ is very nearly the same as for what we call a vacuum, is called the *specific inductive capacity* of the dielectric in question. This ratio is greater than unity for all solid and liquid dielectrics with which we are acquainted. The specific inductive capacity of a perfect conductor would be infinite.

The following very clear statement of the effect produced by changing the dielectric which envelops the parts of a condenser made of two plates, is due to Dr. Hall, and is copied with his permission :

" The fundamental fact concerning static electrical induction as observed by Faraday is this,* that if the two plates of a condenser, separated by air, receive respectively e_1 and $-e_2$ units of electricity when charged to a certain difference of potential, e.g., by connection with the poles of a battery of many cells in series, the same two plates would, if any other medium were substituted for the air, other conditions remaining unchanged, receive respectively Ke_1 and $-Ke_2$ units of electricity, K being some quantity greater than unity. This quantity K is called the specific inductive capacity of the second medium.

" Now, since the difference of potential between A and B is the same in these two cases, the 'electromotive intensity,'† i.e., the force exerted upon unit quantity of electricity, is the same in the two cases at any given point lying in the region through which the change of dielectric extends. If we were to attempt to determine the surface densities of the charges of the conductors by means of the equation ‡

$$\frac{dV_2}{dv} - \frac{dV_1}{dv} + 4\pi\sigma' = 0,$$

the values obtained would be the same for both cases. These would be the actual values of the surface densities if air were used, but would evidently not be the actual surface densities for the other case. For this latter case, the values thus found are called the 'apparent' surface densities, and bear to the true densities the ratio 1 to K.

" We must not conclude from this that A and B with charges Ke_1 and $-Ke_2$ respectively in the second medium would act,

---

* Maxwell's *Treatise on Electricity and Magnetism*, Art. 52.

† Maxwell's *Treatise on Electricity and Magnetism*, Art. 44.

‡ Maxwell's *Treatise on Electricity and Magnetism*, First Edition, Art. 83. See, also, Section 47 of this book.

in *all* electrical respects, like the same bodies with charges e_1 and $-e_2$ in air. Two spheres, A and B, in air, with centres at distance r from each other, and having charges e_1 and $-e_2$ respectively, would attract each other with a force $\frac{e_1 e_2}{r^2}$, whereas the same two spheres with actual charges Ke_1 and $-Ke_2$ in a medium of specific inductive capacity K would attract each other with a force* $\frac{Ke_1 e_2}{r^2}$. This seems at first inconsistent with the fact that the electromotive intensity at any point, as stated above, is the same in both cases. The electromotive intensity at any point, however, means the force that would be exerted upon unit *actual* quantity of electricity at that point, not the force that would be exerted upon unit *apparent* quantity. So the average force exerted by A's charge upon B's charge in either of our two cases is $\frac{e_1}{r^2}$ for each actual unit of B's charge. Hence, the total force exerted by A upon B is $\frac{e_1 e_2}{r^2}$ for the first case, and $\frac{Ke_1 e_2}{r^2}$ for the second case, as stated before."

66. Charge induced on a Sphere by a Charge at an Outside Point. The value at any point P of the potential function due to m_1 units of positive electricity concentrated at a point A_1, and m_2 units of negative electricity concentrated at a point A_2, is

$$V = \frac{m_1}{r_1} - \frac{m_2}{r_2} \quad \text{where} \quad r_1 = A_1 P \text{ and } r_2 = A_2 P.$$

It is easy to see that if m_1 is greater than m_2, so that $m_1 = \lambda m_2$ where $\lambda > 1$, V will be equal to zero all over a certain sphere which surrounds A_2.

If (Fig. 46) we let $A_1 A_2 = a$, $A_1 O = \delta_1$, $A_2 O = \delta_2$, $OD = r$, it is easy to see that

$$\delta_1 = \frac{\lambda^2 a}{\lambda^2 - 1}, \quad \delta_2 = \frac{a}{\kappa^2 - 1}, \quad r^2 = \frac{a^2 \lambda^2}{(\lambda^2 - 1)^2} = \delta_1 \delta_2, \quad \lambda^2 = \frac{\delta_1}{\delta_2},$$

---

* Maxwell's *Treatise on Electricity and Magnetism*, Art. 94.

and
$$a = \frac{\delta_1^2 - r^2}{\delta_1} = \frac{r^2 - \delta_2^2}{\delta_2}.$$
[176]

If PR represents the force f_1 due to the electricity at A_1, and PQ the force f_2 due to the electricity at A_2, the line of action of the resultant force F (represented by PL) must pass through the centre of the sphere, since the surface of the sphere is equipotential.

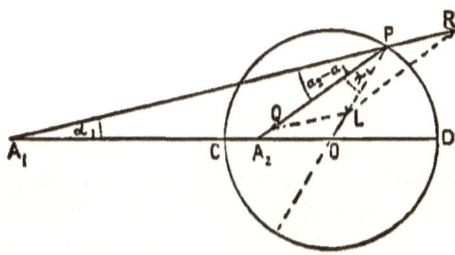

Fig. 46.

The triangles A_1PO and A_2PO are mutually equiangular, for they have a common angle A_1OP, and the sides including that angle are proportional ($r^2 = \delta_1\delta_2$). Hence, from the triangles QPL and A_1PA_2, by the Theorem of Sines,

$$\frac{f_1}{\sin a_1} = \frac{f_2}{\sin a_2} = \frac{F}{\sin(a_2 - a_1)},$$
[177]

$$\frac{r_1}{\sin a_2} = \frac{r_2}{\sin a_1} = \frac{a}{\sin(a_2 - a_1)},$$
[178]

whence
$$F = \frac{af_1}{r_2} = \frac{am_1}{r_2 r_1^2} = \frac{a\lambda m_1}{r_1^3}.$$
[179]

Now, according to Section 50, we may distribute upon the spherical surface just considered a quantity m_2 of negative electricity in such a way that the effect of this distribution at all points outside the sphere shall be equal to the effect of the charge $-m_2$ concentrated at A_2, and the effect at points within the sphere shall be equal and opposite to the effect of the charge m_1 concentrated at A_1. Since F is the force at P in the direction of the interior normal to the sphere, we shall accomplish this if we make the surface density at every point equal to σ, where

$$4\pi\sigma = -F = \frac{-a\lambda m_1}{r_1^3} = \frac{-(\delta_1^2 - r)^2 m_1}{r r_1^3};\qquad [180]$$

and if we now take away the charge at A_2, the value of the potential function throughout the space enclosed by our spherical surface, and upon the surface itself, will be zero. If the spherical surface were made conducting, and were connected with the earth by a fine wire, there would be no change in the charge of the sphere, and we have discovered the amount and the distribution of the electricity induced upon a sphere of radius r, connected with the earth by a fine wire and exposed to the action of a charge of m_1 units of positive electricity concentrated at a point at a distance δ_1 from the centre of the sphere.

If now we break the connection with the earth, and distribute a charge m uniformly over the sphere in addition to the present distribution, the potential function will be constant (although no longer zero) within the sphere, and we have a case of equilibrium, for we have superposed one case of equilibrium (where there is a uniform charge on the sphere and none at A_1) upon another. The whole charge on the sphere is now

$$M = m - m_2 = m - \frac{m_1 r}{\delta_1},$$

and the value of the potential function within it and upon the surface,

$$V = \frac{M}{r} + \frac{m_1}{\delta_1} = \frac{m}{r}.$$

If the conducting sphere were at the beginning insulated and uncharged, we should have $M = 0$, and therefore

$$\sigma = \frac{m_1}{4\pi r}\left(\frac{1}{\delta_1} - \frac{\delta_1^2 - r^2}{r_1^3}\right),\quad\text{and}\quad V = \frac{m_1}{\delta_1}.\qquad [181]$$

If we have given that the conducting sphere, under the influence of the electricity concentrated at A_1 is at potential V_1, we know that its total charge must be $V_1 r - \dfrac{m_1 r}{\delta_1}$, and its surface density

$$= \frac{1}{4\pi r}\left(V_1 - \frac{(\delta_1{}^2 - r^2)\, m_1}{r_1{}^3} \right). \qquad [182]$$

It is easy to see that the sphere and its charge will be attracted toward A_1 with the force

$$\frac{m_1 r}{\delta_1}\left(\frac{m_1 \delta_1{}^2}{(\delta_1{}^2 - r^2)^2} - \frac{V_1}{\delta_1} \right); \qquad [183]$$

and the student should notice that, under certain circumstances. this expression will be *negative* and the force repulsive.

If $m_1 = m_2$, the surface of zero potential is an infinite plane, and our equations give us the charge induced on a conducting plane by a charge at a point outside the plane.

The method of this section enables us to find also the capacity of a condenser composed of two conducting cylindrical surfaces, parallel to each other, but eccentric; for a whole set of the equipotential surfaces due to two parallel, infinite straight lines. charged uniformly with equal quantities per unit of length of opposite kinds of electricity, are eccentric cylindrical surfaces surrounding one of the lines A_2. and leaving the other line A_1 outside. We may therefore choose two of these surfaces, distribute the charge of A_1 on the outer of these, and the charge of A_2 on the inner, by the aid of the principles laid down in Section 50, so as to leave the values of the potential function on these surfaces the same as before. These distributions thus found will remain unchanged if the equipotential surfaces are made conducting.

The reader who wishes to study this method more at length should consult, under the head of Electric Images. the works of Cumming, Maxwell. Mascart. and Watson and Burbury. as well as original papers on the subject by Murphy in the *Philosophical Magazine*, 1833. p. 350. and by Sir W. Thomson in the *Cambridge and Dublin Mathematical Journal* for 1848.

67. The Energy of Charged Conductors. If a conductor of capacity C, removed from the action of all electricity except its own, be charged with M_1 units of electricity, so that it is at potential $V_1 = \dfrac{M_1}{C}$, the amount of work required to bring up to the conductor, little by little, from the walls of the room, the additional charge Δm, is ΔW, which is greater than $V_1 \cdot \Delta M$ or $\dfrac{M_1}{C} \cdot \Delta M$, and less than $(V_1 + \Delta_M V) \cdot \Delta M$ or $\dfrac{M_1 + \Delta M}{C} \cdot \Delta M$.

If the charge be increased from M_1 to M_2 by a constant flow, the amount of work required is evidently

$$\int_{M_1}^{M_2} \frac{M\,dM}{C} = \frac{M_2^2 - M_1^2}{2\,C} = \frac{C}{2}(V_2^2 - V_1^2). \qquad [184]$$

The work required to bring up the charge M to the conductor at first uncharged is then

$$\frac{M^2}{2\,C} = \frac{C V^2}{2} = \frac{M V}{2}. \qquad [185]$$

This is evidently equal to the potential energy of the charged conductor, and this is independent of the method by which the conductor has been charged.

If, now, we have a series of conductors A_1, A_2, A_3, etc., in the presence of each other at potentials V_1, V_2, V_3, etc., and having respectively the charges M_1, M_2, M_3, etc., and if we change all the charges in the ratio of x to 1, we shall have a new state of equilibrium in which the charges are $x M_1, x M_2, x M_3$, etc.; and the values of the potential functions within the conductors are $x V_1, x V_2, x V_3$, etc. The work (ΔW) required to increase the charges in the ratio $x + \Delta x$ instead of in the ratio x is greater than

$$(M_1 \Delta x)(x V_1) + (M_2 \Delta x)(x V_2) + (M_2 \Delta x)(x V_2) + \text{etc.},$$

or $\qquad x\,\Delta x[M_1 V_1 + M_2 V_2 + M_3 V_3 + \text{etc.}]$,

and less than

$$(x + \Delta x)\Delta x\,[M_1 V_1 + M_2 V_2 + M_3 V_3 + \text{etc.}];$$

hence the whole amount of work required to change the ratio from $\frac{x_1}{1}$ to $\frac{x_2}{1}$ is

$$W_2 - W_1 = \frac{x_2^2 - x_1^2}{2}[M_1 V_1 + M_2 V_2 + M_3 V_3 + \text{etc.}]. \quad [186]$$

If in this equation we put $x_1 = 0$ and $x_2 = 1$, we get for the work required to charge the conductor from the neutral state to potentials V_1, V_2, V_3,

$$W = \tfrac{1}{2}[M_1 V_1 + M_2 V_2 + M_3 V_3 + \cdots] = \tfrac{1}{2}\sum(MV). \quad [187]$$

68. If a series of conductors A_1, A_2, A_3, etc., are far enough apart not to be exposed to inductive action from one another, and have capacities C_1, C_2, C_3, etc., and charges M_1, M_2, M_3, etc., so as to be at potentials V_1, V_2, V_3, etc., where $M_1 = C_1 V_1$, $M_2 = C_2 V_2$, $M_3 = C_3 V_3$, etc., we may connect them together by means of fine wires whose capacities we may neglect, and thus obtain a single conductor of capacity

$$C_1 + C_2 + C_3 + \cdots = \sum(C).$$

The charge on this composite conductor is evidently

$$M_1 + M_2 + M_3 + \cdots = \sum(M);$$

and if we call the value of the potential function within it V, we shall have

$$V \cdot \sum(C) = \sum(M);$$

whence

$$V = \frac{C_1 V_1 + C_2 V_2 + C_3 V_3 + \cdots}{C_1 + C_2 + C_3 + \cdots}, \quad [188]$$

a formula obtained, it is to be noticed, on the assumption that the conductors do not influence each other.

The energy of the separate charged conductors before being connected together was

$$W = \tfrac{1}{2}(M_1 V_1 + M_2 V_2 + M_3 V_3 + \cdots) = \tfrac{1}{2}\left(\frac{M_1^2}{C_1} + \frac{M_2^2}{C_2} + \frac{M_3^2}{C_3} + \cdots\right)$$

$$= \tfrac{1}{2}\sum\left(\frac{M^2}{C}\right). \quad [189]$$

and the energy of the composite conductor is

$$W' = \frac{\frac{1}{2}(M_1 + M_2 + M_3 + \cdots)(C_1 V_1 + C_2 V_2 + C_3 V_3 + \cdots)}{C_1 + C_2 + C_3 + \cdots}$$

$$= \frac{\frac{1}{2}(M_1 + M_2 + M_3 + \cdots)^2}{C_1 + C_2 + C_3 + \cdots} = \frac{\frac{1}{2}\left[\sum (M)\right]^2}{\sum (C)}, \qquad [190]$$

which is always less than E unless the separate conductors were all at the same potential in the beginning.

EXAMPLES.

1. Show that in general the surface density of a charge distributed on a conductor is greatest at points where the convex curvature of the surface of the conductor is greatest.

2. A hollow in a conductor is at the uniform potential V_1 when a charge is communicated to a conductor within the cavity sufficient to raise this conductor to potential V_2 if it were in empty space. Give some idea of the changes brought about by this charge.

3. Show that a field of electric force consists wholly of non-conductors bounded, if at all, by conducting surfaces.

4. Prove that if a distribution of electricity over a closed surface produce a force at every point of the surface perpendicular to it, this distribution will produce a potential function constant within the surface.

5. Two conducting spheres of radii 6 and 8 respectively are connected by a long fine wire, and are supposed not to be exposed to each other's influences. If a charge of 70 units of electricity be given to the composite conductors. show that 30 units will go to charge the smaller sphere and 40 units to the larger sphere, if we neglect the capacity of the wire. Show also that the tension in the case of the smaller sphere is $\dfrac{25}{288\pi}$ per square unit of surface.

6. An uncharged sphere A, of radius r, occupies the centre of the otherwise empty, equipotential cavity, enclosed by a spherical shell B of radii r_i and r_o, so large that the effect inside the cavity of the charge induced on B by a charge m, communicated to A from without, may be neglected. If the value of the potential function within the cavity before A was charged was C, at what potential is A now? Ans. $C + \dfrac{M}{r}$.

7. The first of three conducting spheres, A, B, and C, of radii 3, 2, and 1 respectively, remote from one another, is charged to potential 9. If A be connected with B for an instant, by means of a fine wire, and if then B be connected with C in the same way, C's charge will be $3 \cdot 6$. [Stone.] If, in the last example, all three conductors be connected at the same time, C's charge will be $4 \cdot 5$.

8. A charge of M units of electricity is communicated to a composite conductor made up of two widely-separated ellipsoidal conductors, of semiaxes 2, 3, 4 and 4, 6, 8 respectively, connected by a fine wire. Show that the charges on the two ellipsoids will be $\dfrac{1}{3} M$ and $\dfrac{4}{3} M$ respectively. [Stone.]

9. Can two electrified bodies repel each other when no lines of force can be drawn from one body to the other?

10. Two conductors, A and B, connected with the earth are exposed to the inductive action of a third charged body. Do A and B act upon each other? If so, how?

11. Show that two equal conductors similarly placed with respect to each other always repel each other if raised to the same potentials and insulated; but that if the volume of the potential function within the conductors differ never so little from each other, they will repel each other at great distances, but at very near distances (supposing no spark to pass) they will attract each other. [Cummings.]

12. The superficial density has the same sign at all points of a conducting surface outside which there is no free electricity.

13. Show that $r \div \delta$ of the unit tubes of force proceeding

from an electrified particle, at a distance δ from the centre of a
conducting sphere of radius r, which is put to earth, meet the
sphere if there are no other conductors in the neighborhood,
and that the rest go off to "infinity."

14. A charged insulated conductor A is so surrounded by a
number of separate conductors B, C, D, ···, which are put to
earth, that no straight line can be drawn from any point of A
to the walls of the room without encountering one of these other
conductors: will there be any induced charge on the walls of
the room? See Section 37.

15. Two uniform straight wires of equal density, each two
inches long, lie separated by an interval of one inch in the
same straight line. Find the equation of the equipotential surfaces due to these wires, and find what must be the density of a
superficial distribution of matter on one of these surfaces which
at all outside points would exert the same attraction as the
wires do.

16. An insulated conducting sphere of radius r charged with
m units of positive electricity is influenced by m units of positive electricity concentrated at a point 2r distant from the centre of the sphere. Give approximately the general shape of the
equipotential surfaces in the neighborhood of the sphere.

Give an instance of a positively electrified body whose potential is negative.

17. A conductor, the equation of whose surface is

$$\frac{x^2}{25} + \frac{y^2}{16} + \frac{z^2}{9} = 1,$$

is charged with 80 units of electricity; what is the density at a
point for which $x = 3$, $y = 3$?

If the density at this point be a, what is the whole charge on
the ellipsoid?

18. Prove that the capacity of n equal spherical condensers
when arranged in cascade is only about $\frac{1}{n}$th of the capacity of
one of the condensers; but that if the inner spheres of all the

condensers be connected together by fine wires, and the outer conductors be also connected together, the capacity of the complex condenser thus found is about n times that of a single one of the condensers.

19. Prove that if the charges of a system of conductors be increased. the increment of the energy of the system is equal to half the sum of the products of the increase in the charge first conducted into the sum of the values of the potential function within it at the beginning and the end of the process, or to half the sum of the products of the increment of the value of the potential function in each conductor into the sum of the original and final charges on that conductor. [Maxwell.]

20. Prove that if the charges of a fixed system of conductors be increased, the sum of the products of the original charge and the final potential of each conductor is equal to the sum of the products of the final charge and the original potential. [Maxwell.]

21. Discuss the following passage from Maxwell's *Elementary Treatise on Electricity* :

" Let it be required to determine the equipotential surfaces due to the electrification of the conductor C placed on an insulating stand. Connect the conductor with one electrode of the electroscope, the other being connected with the earth. Electrify the exploring sphere.* and, carrying it by the insulating handle, bring its centre to a given point. Connect the electrodes for an instant, and then move the sphere in such a path that the indication of the electroscope remains zero. This path will lie on an equipotential surface."

22. Prove that the coefficients of potential (p) and induction (q) treated in Article 59 have the following properties :

(1) The order of the suffixes of any p or any q can be inverted without altering the value of the coefficient. or, in other words.

$$p_{u} = p_{u}. \quad \text{and} \quad q_{v} = q_{u}.$$

---

* A very small conducting sphere fitted with an insulating handle.

(2) All the p's are positive, but p_{kl} is less than either p_{ll} or p_{kk}.

(3) Those q's whose two suffixes are the same are positive; the others are negative. That is, q_{kk} and q_{ll} are positive; but q_{kl} is negative and is, moreover, numerically less than either of the others.

23. Prove that the following theorems (Maxwell's *Elementary Treatise on Electricity*) are contained in the statements of the preceding problem:

(1) In a system of fixed insulated conductors, the potential function in A_k due to a charge communicated to A_l is equal to the potential function in A_l due to an equal charge in A_k.

(2) In a system of fixed conductors connected, all but one, with the walls of the room, the charge induced on A_k when A_l is raised to a given potential is equal to the charge induced on A_l when A_k is raised to an equal potential.

(3) If in a system of fixed conductors, insulated and originally without charge, a charge be communicated to A_k which raises it to potential unity and A_l to potential n, then if in the same system of conductors a charge unity be communicated to A_l, and A_k be connected with the earth, the charge induced on A_k will be $-n$.

24. A condenser consists of a sphere A of radius 100 surrounded by a concentric shell whose inner radius is 101 and outer radius 150. The shell is put to earth, and the sphere has a charge of 200 units of positive electricity. A sphere B of radius 100 outside the condenser can be connected with the condenser's sphere by means of a fine insulated wire passing through a small hole in the shell. B is connected with A; the connection is then broken, and B is discharged; the connection is then made and broken as before, and B is again discharged. After this process has been gone through with five times, what is A's potential? What would it become if the shell were to be removed without touching A?

25. Suppose the condenser mentioned in the last problem insulated and a charge of 100 units of positive electricity given to

the shell. What will be the potential of the sphere? of the shell? If we then connect the sphere with the earth by a fine insulated wire passing through the shell, what will the charge on the shell be? What will be the potential of the shell? If next *A* be insulated, and the shell be put to earth, what will be *A*'s potential? What will be its potential if the shell be now wholly removed?

26. A spherical conductor of radius r is surrounded by a concentric conducting spherical shell of radii R_1 and R_0, and the outer surface of this shell is put to earth. If the inner conductor be charged, show the effect at all points in space of moving the conductor so that it shall be eccentric with the shell. How is the capacity of the system changed by this?

27. Prove that if the spherical surfaces of radii a and b, which form a spherical condenser, are made slightly eccentric, c being the distance between their centres, the change of electrification at any point of either surface is $\dfrac{3\,abc \cdot \cos\theta}{4\pi\,(b-a)\,(b^3-a^3)}$, where θ is the angular distance of the point from the line of centres, and where the difference between the values of the potential function on the two surfaces is unity.

28. Show that if an insulated conducting sphere of radius a be placed in a region of uniform force (X), acting parallel to the axis of x, the function $-X \cdot x\left[1 - \dfrac{a^3}{r^3}\right] + C$ satisfies all the conditions which the potential function outside the sphere must satisfy, and is therefore itself the potential function. Show that the surface density of the charge on the sphere is $\dfrac{3\,x\,X}{4\pi a}$.
[Watson and Burbury.]

MISCELLANEOUS PROBLEMS.

1. Prove that the attraction due to a homogeneous hemisphere of radius r is zero, at a point in the axis of the hemisphere distant $\frac{3}{7}r$ approximately from the centre of the base.

2. Show that the attraction at the origin due to the homogeneous solid bounded by the surface obtained by revolving one loop of the curve $r^2 = a^2 \cdot \cos 2\theta$, is $\frac{1}{8}\pi a$.

3. If the earth be considered as a homogeneous sphere of radius r, and if the force of gravity at its surface be g, show that from a point without the earth, at which the attraction is $\frac{n-1}{n}g$, the area $2\pi r^2 \left(1 - \frac{n-1}{n}\right)$ on the surface of the earth will be visible.

4. A spherical conductor A, of radius a, charged with M units of electricity, is surrounded by n conducting spherical shells concentric with it. Each shell is of thickness a, and is separated from its neighbors by empty spaces of thickness a. Show that the limit approached by V_A as n is made larger and larger is $\frac{M}{a}$ (nat. log 2), and that for a finite number of shells

$$V_A = \frac{M}{a} \int_0^1 \frac{1 + x^{2n+1}}{1+x} \, dx. \quad \text{[Stone.]}$$

5. If two systems of matter (M and M'), both shut in by a closed surface S, give rise to potential functions (V and V'), which have equal values at every point of S, whether or not S is an equipotential surface of either system, then V cannot differ from V' at any point outside S, and the algebraic sum of the matter of either system is equal to that of the other. [See Section 52, and Watson and Burbury's *Mathematical Theory of Electricity and Magnetism*, § 60.]

6. Show that if two distributions of matter have in common an equipotential surface which surrounds them both, all their equipotential surfaces outside this will be common.

7. Prove that if V be the potential function due to any distribution of matter over a closed surface S, and if σ' be the density of a superficial distribution on S, which gives rise to the same value of the potential function at each point of S as that of a unit of matter concentrated at any given point O, then the value at O of the potential function due to the first distribution is $\int V \cdot \sigma' \cdot dS$.

www.ingramcontent.com/pod-product-compliance
Lightning Source LLC
Chambersburg PA
CBHW030901050726
47500CB00009B/902